Project

APOLLO

Exploring the Moon

by Robert Godwin

CONTENTS

All rights reserved under article two of the Berne Copyright Convention (1971).
We acknowledge the financial support of the Government of Canada through the Book
Publishing Industry Development Program for our publishing activities.

Published by Apogee Books, Box 62034, Burlington,
Ontario, Canada, L7R 4K2, http://www.apogeebooks.com
Tel: 905 637 5737
Printed and bound in Canada
Project Apollo Exploring the Moon by Robert Godwin
ISBN 1-894959-37-X
ISBN13 978-1894959-37-7
©2006 Robert Godwin

EXPLORING THE MOON

It could easily be argued that the extraordinary journey leading us to Tranquillity Base began when the first set of human eyes gazed at the quick-silver moon and pondered its purpose. That moment is lost in the depths of pre-history and it is only with the birth of human record keeping, the historic era, that we can state with any certainty, who amongst the pantheon of genius finally set us on the path to another world.

There were the philosophers of ancient times who first believed that the moon was a destination; a goal that could actually be attained. They included Thales and Heraclitus, Plutarch and Ptolemy, Aristarchus and many others. These titans of early science spent entire lifetimes, bantering back and forth, and delving into the nature of the universe. They had their counterparts in other cultures but so few of those left their musings in written form for future generations to study that we are mainly left with the Greeks.

Then there was the intellectual revival of the 16th century when Copernicus, Tycho and Kepler challenged centuries of dogma to try and establish the truth about the heavens, some risking their own well-being to unlock the mysteries of the universe. They rediscovered old truths and passed on their discoveries to the experimenters and scientists of the 17th century; immortals like Galileo and Newton, who used new technologies and their own brilliant minds to shatter the preconceptions of millennia and expose the glory of our cosmos with unprecedented accuracy. The distinction between science and philosophy was becoming more clearly defined and proponents of the latter discipline, like Voltaire, used this new found knowledge with surgical precision in a final assault on the politically motivated contrivances of the ruling elite. The new astronomy was proving to be a powerful weapon in the war to remove the divide between rulers and minions.

Then as books became more prevalent, and people became more educated, the story fell into the purview of the novelist. These mavericks were at the end of a long line of story-tellers, a line with a heritage that trailed back into the haze of pre-history, but now they would venture into the realms of outer space. In the 19th century a revolution in industry encouraged people to think that anything might be possible, including actual flights to the moon. Verne and Wells and Lasswitz and Poe would ensure their own immortality by directly inspiring the generation that would actually build flying machines. The 19th century had distinguished itself with photography and steel, steam engines and electricity, rubber tires and cement, the automobile and pasteurization, but the 20th century would truly earn the reputation of being an unparalleled era, a century of computers and aircraft, of telecommunications and weather fore-casting, of genetic engineering and nuclear power, of antibiotics and heart transplants, television and, of course, space travel.

This complex framework of technology is so thoroughly compounded it is all but impossible to isolate one event without accounting for innumerable others. Space travel would not have been possible without the endless chain of contributors dating back to the beginning of history, but there are clearly defined moments in the record when someone, or something, intervenes and provides the impetus to propel us to unforeseen heights. Such was the case with America's *Apollo* program.

As has been outlined in this book's companion volume (*Project Apollo – The Test Program*) the moon landings of the 1960's and 70's would not have been possible without an improbable chain of events culminating in the exhortation of a young President, calling his nation to greatness. Long after many of the 20th century's great litany of achievements are forgotten, people will remember Tranquillity Base and they will recall the names of the heroes who risked their lives to make their species the first in four billion years to venture into the cosmos. In a very, very long list of human achievements, Apollo is extraordinary.

On July 16th 1969 launch vehicle Saturn/Apollo 506 pronounced its departure from planet Earth with a deafening roar thundering across the Florida everglades. It has been estimated that anything from 400,000 to over a million souls stood and watched the spectacle with their own eyes. Some stood in silent awe at the power unleashed by the most powerful controlled explosion in the history of humanity, but considerably more of them shouted and cheered, waving their arms in the air and egging it on as though their combined enthusiasm could overrule any possible shortcomings in the billowing leviathan. For those who remember that day it was an unforgettable and unprecedented few minutes where the hopes and dreams of an entire species rode along with three bold and courageous explorers. Our hearts and minds were completely overrun by the aspirations of America's best and brightest. Although in years to come Apollo would be reviled by pundits as nothing more than a stunt to outdo the communists, there were millions of people around the world who perceived it as a beacon of optimism, the first glimmerings that humanity might be capable of something truly profound.

On the evening of July 20th 1969 (American eastern standard time) Neil Armstrong and Buzz Aldrin would don their multi-layered white protective garments and step outside to, as Neil Armstrong would later put it, "play in the sandbox." For 140 minutes they lived entirely on another world, doing what humans do, running and digging, chatting with friends back home and setting up ingenious experiments. They proclaimed proudly that they had come to this remote forbidding world "In Peace, for all mankind," and then they came home. (For further details about Apollo 11 see the companion volume *Apollo 11 – First Men on the Moon*.)

There were those who had only supported this most incredible adventure because they came from a different era, one of war and hardship. They were terrified that their barking and belligerent counterparts in Russia, and elsewhere, might somehow conquer the world, and so they had grudgingly sponsored this most ambitious program in an effort to win a war of psychology. It is undeniable that for those who were thus motivated Apollo was a resounding success. No longer were the citizens of other nations so sure that communism might prove itself a better system of government. American science and the democratic system were overwhelmingly and convincingly superior. Never had a flight carried so much baggage.

The repercussions from this incredible war of doctrines are still being felt today. In the same way as the civilization that had built the pyramids in Egypt would make its presence known around the ancient world for centuries to come, the enormous technological benefits of the Apollo program will stretch into the future for untold generations.

Our understanding of the universe was subsequently given an unprecedented push forward by the next five moon landings. After the successful return of Armstrong, Aldrin and Collins the remainder of NASA's dauntless Apollo astronauts prepared to follow behind them. At this point it could be argued that the battle had been won and there were no compelling reasons to risk more lives but it would have been impossible to simply build one moon rocket and one spacecraft. To perfect such an incredibly complex system of machinery had required many test flights and prudence dictated that there also be many back-up vehicles. No one had known for sure if Armstrong would be able to successfully land his fragile lunar module and, right up until he announced their arrival, the team at NASA had not known for any certainty that Apollo 11 would be the first to succeed. So there were many redundancies, including a half dozen more fully trained crews and the equipment necessary to get them to the moon.

On 14th November 1969 at 11:22 a.m. an all Navy crew of Charles "Pete" Conrad, Richard Gordon and Alan Bean took their moment in the kerosene and oxygen limelight as their 365 foot tall Saturn V clawed its way upwards from the launch pad at the Kennedy Space Center. Only thirty six seconds into the flight all hell broke loose as the giant rocket was struck by the first of two lightning strikes. At that moment the entire computer platform flat-lined and the crew were plunged into darkness. By any normal standards Pete Conrad might have been expected to engage the Abort system but as he later noted that "baby just kept chugging along" so he left well enough alone. Moments after a second lightning discharge, someone on the ground relayed to the Capsule Communicator the necessary procedure to reboot the flight computers and legend would have it that rookie Alan Bean was the only one on board who knew what was meant by the cryptic instructions dictated from the ground.

After saving the day Al Bean and his compatriots continued onwards on what became known as a "hybrid" trajectory. This unusual flight path would mean that the spacecraft would be steered away from the previously used "free return" trajectories used by Apollos 8, 10 and 11. A free return meant that as the spacecraft left Earth orbit it would head onto a path that, left uninterrupted, would use the moon's gravity to slingshot the Apollo back to Earth by simply looping around the back of the moon. It was a more-or-less risk-free course that would bring crews home even if their main engines failed. Thirty hours and fifty two minutes into their voyage Apollo 12's crew would veer away from this safe route and put their fate in the hands of the continued dependability of their engines. Those same engines would now be their only means to return to Earth.

Once Apollo 12 was in lunar orbit it was determined that it was further off course than even Apollo 11 had been, but it was a small enough problem that the error could be corrected during descent. Part of this error was due to the landing site not being where it was expected to be! It was not a huge difference but combined with two other minor problems it was enough to require adjustments. Because the landing site was further west there was more time to update the computer than on Apollo 11, and so these updates began soon after powered descent. Two other precautions were taken to avoid the problems that had been presented during Apollo 11's descent. Firstly the undocking took place in a position where the two craft were oriented radially in relation to the moon and was done much more delicately to avoid any unwanted accelerations. Secondly the LM began its descent already in the face up position. This precluded any possibility of getting pretty pictures of the moon during the early part of the descent (Buzz Aldrin on Apollo 11 had turned on the film camera before the beginning of powered descent and filmed all the way through touchdown, about 14 minutes). Bean would not engage the film camera until only seven minutes before touchdown because there was effectively nothing to see but black sky prior to the gradual pitchover of *Intrepid*.

At approximately 1:45 pm (EST) in the afternoon of November 19th Pete Conrad and Al Bean ignited the descent engine on the Lunar Module *Intrepid*. The minimal thrust of about 1300 pounds poured out of the engine bell and 27 seconds later the computer engaged full throttle, increasing the thrust to its nominal rating of 9,900 pounds. Fifty eight seconds later the crew punched in the revised landing site data (a discrepancy from the predicted location of 4200 feet) and they were on their way.

As they continued through their descent, the computer started to report the location of the landing site as it should have appeared through a simple device in the LM window called a Landing Point Designator (LPD) (a thin line on the glass with numbers scaled along its length). As the computer displayed a number, Alan Bean would read that number back and Conrad could then compare that number against what he was seeing outside—behind the lines on the window. The first number the computer displayed told them that *Intrepid* was heading dangerously close to the rim of their target crater, about four hundred feet north of where they had expected to be. The next LPD number placed them four hundred feet further down-range and slap bang in the center of a four hundred foot wide hole in the ground called *Head* crater. As the approach vector was gradually improved the LPD moved another four hundred feet west before gradually backtracking eight hundred feet east and back to the edge of Surveyor crater where Conrad finally and gently set down the fragile lander. The place where the second humans touched down on the moon was just under eight hundred feet from the computer's target and about four hundred feet from the *Surveyor* spacecraft that had been sent there thirty months earlier.

Once on the surface Conrad and Bean prepared for their EVA (Extra-Vehicular Activity or, in plain English, moonwalk). This began somewhat late because they had taken some extra time to try and determine their exact position on the surface. The proximity of the lunar horizon and the lack of atmosphere made it difficult to gauge distances and it would not be until Conrad made his way outside that he could be entirely certain of their position, an extremely important need-to-know fact if they were to successfully explore and then later rendezvous with Gordon in the CSM. Conrad and Bean would spend nearly eight hours on the lunar surface—spread over two four hour periods—with a 12 hour rest period between. The visual record of their exploits is spotty since the television camera failed just about 50 minutes into the first EVA. The two astronauts had not trained with the real hardware prior to the flight and the delicate device was accidentally pointed directly into an intense solar reflection coming off the lunar module's protective foil.

On the first EVA the crew concentrated on setting up the first complete lunar experiment package known by its acronym, ALSEP. This compact, yet effective bundle of equipment was to be powered by a small nuclear power plant, the first of its kind to be deployed on the moon. In a moment of bizarre comedy Alan Bean was forced to use a hammer to successfully extract the fuel rod for the power plant, it was a technique he had earlier applied in an attempt to restore the television camera. Although the method failed on the camera it worked on the fuel rod for the SNAP-27 (System for Nuclear Auxiliary Power) which was essentially an atomic battery.

On the first day Conrad and Bean deployed the ALSEP about 450 feet north west of the LM and then ventured out about another 600 feet to the edge of a large old crater called *Middle Crescent*. On the second day they undertook a more ambitious hike which led them first back to the ALSEP, then south west around the rim of *Head* crater before turning almost due south to a small crater called *Bench*. They then reached their furthest point from the LM when they turned due west and walked about 100 yards to a small, relatively new, crater called *Sharp-Apollo* before backtracking due east past *Bench* and then on, another 200 yards, to *Halo* crater; a very small crater about 200 yards due south of the LM. Finally they turned north by north-east and worked their way around their target crater, named after its occupant, the *Surveyor 3* robot lander.

On the far eastern rim of the 200 yard wide crater the Surveyor was settled on a 12 degree slope and was now bathed in the morning sunlight (on the first day it was still concealed in shadow.) Conrad and Bean set about documenting the area around the robot, taking samples of rocks that had been photographed two years earlier by the robot, for comparison purposes. They chopped off the television camera and sampling scoop and they noted that the vehicle was covered in a fine layer of dust, mainly on the outwardly exposed surfaces, and that the electrical insulation was degraded to the texture of old asbestos. After completing some spectacular photographs the two explorers returned to the *Intrepid* and prepared to return home.

The flight of Apollo 12 represented a logical next step in the NASA lunar program. The key objectives were all accomplished. They had landed precisely where they wanted to, they had deployed a complete experiment package, they had successfully utilized the new hybrid trajectory and they had even survived what should have been a show-stopping lightning strike. The team at NASA were emboldened to continue with their increasingly ambitious exploration program. Meanwhile, the forces in Washington were already compelling NASA to tighten its belt. Just seven weeks after Apollo 12 returned home safely, NASA cancelled Apollo 20, a flight that was to have landed at the enormous crater Tycho. Apollo 12 commander Pete Conrad would probably have commanded the mission, along with future Skylab astronauts Paul Weitz and Jack Lousma. Just three months later Apollo 13 began its hapless journey towards an area on the moon called Fra Mauro.

Apollo 13 was commanded by long time veteran Jim Lovell who had already seen the moon from up-close on Apollo 8. Lovell would be bringing along a rookie crew comprising of Fred Haise as his LM pilot and at the last minute his Command Module pilot, Thomas Mattingly, would be replaced by Jack Swigert. Fra Mauro was the first landing site that was not located in one of the dark Mare regions of the moon, it was a slightly hilly area located about six degrees east of the Apollo 12 landing site. Of the 26 objectives laid out for Apollo 13 only two were destined to be completed, and both of those ironically involved a crash on the moon. (See details later in this book).

The launch of Apollo 13 took place on April 11th 1970 at 2.13 pm EST (some keen eyed observer noted that this was 13.13 hours at Mission Control in Houston.). During the launch the second stage lost an engine early but the tremendous redundancy of the Saturn compensated for this and the third stage was able to burn an additional nine seconds to attain the desired velocity. Three hours after launch the Command Module, named *Odyssey*, pulled clear of the SLA and turned to dock with the Lunar Module *Aquarius*. The third stage of the Saturn then made an evasive maneuver to avoid the twin spacecraft and was set on a collision course with the moon. 74 hours later it crashed into the lunar surface with a force of 11.5 tons of TNT, setting off the Apollo 12 seismometer and thus fulfilling one mission objective, but by then almost no one was watching. Another much smaller and insidious explosion had taken place elsewhere.

About twenty hours earlier a piece of melted wiring inside one of the oxygen tanks received a command to send power to a fan. The wiring was melted because a contractor had not implemented a directive to change the specifications inside the tank and then someone had applied the wrong voltage. The insulation couldn't take the heat and the contacts welded shut. When CM pilot Jack Swigert threw a switch to stir up the oxygen tank with the small internal fan, a fire started which, incredibly, took over 90 seconds to propagate into an all-out explosion. Unfortunately, all this happened about 25 hours after Apollo 13 had changed course onto a hybrid trajectory.

Apollo 13 was only half-way to the moon but it was all-but impossible to simply turn around, the only hope to get back quickly was to keep going forward. Five and a half hours after the explosion the crew were compelled to try and get back onto a free-return course by using the descent engine of the lunar module. This risky procedure was induced by the apparent loss of the main service module engine; which may have been disabled along with most of the spacecraft's environmental system. Once the crippled Apollo 13 went around the back of the moon the LM engine was fired again with the intention of speeding up the return journey. Each time one of these burns took place the crew were put into situations that had barely been conceived, much less simulated. While all of this strategic maneuvering was taking place the cabin of the Command Module had been powered down to save the precious batteries necessary for re-entry. This had some adverse effects, firstly the cabin began to fill with condensation and in zero-g that meant water was getting into places like the control panels, secondly it got very cold and the crew suffered considerably, to the point that Fred Haise became quite ill. Then there was the lack of air. Because the explosion had surgically removed most of the air supply the three men had to use the lunar module's considerably smaller supply. This was then compounded by an increase in poisonous carbon dioxide because the air scrubbers in the LM couldn't handle the extra load. The ground controllers jury-rigged an impressive solution that involved a plastic bag, a piece of cardboard and a sock.

At 105 hours and 19 minutes into the mission the LM engine was fired again, for fifteen seconds, to raise the entry flight path angle, and then thirty two hours later the LM's attitude control system performed the final precision maneuvers to place *Odyssey* on course for the Pacific ocean. Thirty minutes later the service module was jettisoned and for the first time the three men saw just how close they had come to oblivion. One entire side panel had been excised from the side of the cylindrical craft, exposing the damage within. The ship's fuel cells had been assailed from beneath by the huge pulse of gas from the disintegrating oxygen tank. Amazingly, because the explosion had gone upwards, the bulkhead between the damaged section and the adjacent propellant tanks had not been breached. Undoubtedly, things could have been a lot worse for Apollo 13.

Three and a half hours later the crew had moved everything they needed into the command module and prepared it for reentry. The lunar module *Aquarius*, with all of its stages still intact, was ejected towards a fiery appointment with the Earth's atmosphere. For the last hour and eleven minutes the three intrepid astronauts sat and waited to see if their heat shield had been compromised by the explosion of four days earlier. Fortunately all was well and 142 hours 54 minutes and 41 seconds after leaving the Earth, Apollo 13 returned with its crew in one piece. It was an unprecedented feat and a remarkable tribute to the NASA team; but it would have long-lasting repercussions.

Less than two months later the eager beavers in Washington were in the middle of a typically interminable display of navel-gazing. The taxpayers' dollars were consumed at a dizzying pace by the committee assigned to investigate what had gone wrong. Gathered in amongst the expert witnesses, such as the crew and their peers, were the usual array of badly prepared bureaucrats, attempting to look good while examining a problem that was clearly above the heads of many of them. Once the accident had been explained, little time was spared before it was decided to eradicate two more Apollo lunar missions. Only a dozen weeks after the publication of the Congressional analysis it was decided that Apollo 14 would go to Fra Mauro and complete 13's mission, Apollo 15 would be given Apollo 19's destination at Hadley, Apollo 16 would still go to Descartes and Apollo 17 would be reassigned to Apollo 14's target in Taurus-Littrow. Thus, two more Saturn V's, along with their lunar modules, were consigned to the scrap-heap with the stroke of a pen. Plans to go to Censorinus, Copernicus, Tycho and the Marius Hills were abandoned, all in less than 9 months.

Five months after this brutal gutting of the Apollo lunar program a heavily modified Saturn V was dispatched to launch complex 39, it was carrying Apollo 14 and it would be manned by America's oldest and least experienced veteran, Alan Shepard. In 1961 Shepard had blazed America's space trail with a fifteen minute sub-orbital flight, but he hadn't flown since. This was mainly due to an ear disorder for which he had been grounded. Now that the flight roster had been completely dismantled and flights reassigned, Shepard found himself on his way to Fra Mauro, along with Edgar Mitchell and accompanied most of the way by Stuart Roosa.

A remarkable amount of work had been done on the Apollo spacecraft in the eight months since Apollo 13 (a list of these changes can be seen further on in this book). NASA management were now relatively comfortable about making another assault on the high ground. On 31st January 1971 at 4.03 pm EST, after some delays, the Saturn/Apollo vehicle 509 plowed its way upwards and onto a 75.5 degree azimuth en route to Earth orbit. About three hours later trouble struck when the command/service module, piloted by Stu Roosa, couldn't dock with the lunar module. A worst case scenario was quickly brewing in which another Apollo landing would have to be scrubbed. For nearly two hours the crew and technicians on the ground applied an assortment of fixes and finally on the sixth attempt the two vehicles finally locked together. After ensuring that there were no leaks, the hardware was removed from the tunnel between the two craft and inspected, live television pictures were sent to the ground but nothing could be found wrong with the system.

After a late launch and then the subsequent docking problems, the crew of Apollo 14 might have been forgiven for thinking that the rest of the flight might be problem-free. However, during the LM's powered descent an emergency warning light suggested that a particle of loose solder might be floating behind the control panel. If the crew were to descend with this arrangement it might cause an unplanned abort and ultimately endanger their lives.

The programmers at MIT were obliged to hastily write an update to the LM software. Mitchell and Shepard punched in the patch and proceeded unhindered towards Fra Mauro. As if that problem had not been enough, moments later the landing radar refused to engage. With no conventional frame of reference a landing was thought to be impossible without the radar, consequently an abort was the accepted procedure. To this day it is a subject of contentious debate whether Shepard would have flown in manually had the radar not been fixed. Fortunately he never had to make the decision when the breaker was recycled and the radar locked-on. With all of these problems it is quite remarkable that Shepard and Mitchell were able to land within 60 feet of their desired target.

Once on the surface the two astronauts went about setting up the largest science package yet deployed. It incorporated modified versions of atmospheric, seismological and passive experiments. One of these, the Active Seismic Experiment (ASE), incorporated a mortar with four rocket-propelled grenades. This extraordinary device was supposed to fire a sequence of explosive devices outwards from the science station to distances ranging from 500 to 5000 feet. It was originally intended that the device would not be activated until the crew were safely back in orbit. Any chance of dramatic TV pictures were non-existent as the TV camera was still reliant on the LM. However it turns out that the assembly was never fired as it was later determined that the grenade exhaust might contaminate the other experiments. Anyone with the right frequency and codes could probably set it off today. Edgar Mitchell later recalled that he and Shepard were inclined to "stay away" from the launcher. The ASE was not a total failure though as the experiment also called for Mitchell to fire a "thumper" into the ground using charges similar to shotgun shells. About 25% of these didn't fire but useful results were returned from Mitchell's efforts. A geophone assembly was installed across the center of a crater pair called *Doublet* just over 300 feet south west of the LM while the rest of the ALSEP was set up about four hundred feet due west of the landing site.

On their second EVA, Mitchell and Shepard would attempt to trek a round trip of almost two kilometers, to the summit of *Cone* crater. To allow such a long-range and risky hike the Apollo space-suits had undergone a series of upgrades. One of the new devices was the Buddy Secondary Life Support System (BSLSS). The BSLSS would be mounted to another new addition—the Modularized Equipment Transporter (MET). The BSLSS would allow the backpacks to be lighter and, similar to common scuba-diving equipment, would allow one astronaut to assist the other by sharing coolant and oxygen. The MET was the real bonus on this EVA. This small rickshaw-type trolley was the key to conducting useful science once the astronauts arrived at their destination. Although it had been designed with the moon in mind it was found to bounce around due to its lunar-weight being only 25 pounds. Shepard and Mitchell took to walking in single file so that they could recover any errant equipment which might bounce off the cart as they proceeded through the rugged terrain around *Cone*. That same ruggedness ultimately foiled the moonwalkers from finding their target, although it was later determined that when they turned back they had come within a mere 75 feet of their destination. Such are the problems of lunar exploration. Large boulders from the ancient impact which had formed *Cone* were photographed and sampled by Shepard and Mitchell. On the way out and back they examined rocks extensively around a cluster of craters called *Triplet* and *Weird*. Overall they walked just over a mile (about 1.8 km) in their journey.

The two explorers returned to the LM in time to conduct one more critical experiment—Al Shepard's legendary study of the flight dynamics of a golf-ball in one-sixth gravity. Although the anecdotes abound that the ball flew clean across the horizon, a revealing photograph later taken from the LM showed that it had flown only fifty feet or so. Enough at least to keep the legend alive and to keep people wondering how long it will be before the next person will yell "Fore!" in the lunar sand-traps.

The crew of Apollo 14 returned safely to the Earth with over 100 pounds of lunar rocks. Stuart Roosa's photography was exquisite (although sadly one film magazine had been left on the surface by his crewmates). The science station began to function perfectly and added more data to the scientists' already burgeoning files.

After Apollo 14 a brash decision was taken to push the Apollo/Saturn hardware to its design limitations for the last three missions. Apollo 15 was capable of taking a full two tons of additional useful equipment into space. This was achieved by narrowing the margins in every conceivable manner. For example the launch window was deliberately placed in the summer since this reduced the inevitable winter wind factor at launch time. The Saturn V had its reserve propellant margins reduced and was launched into a slighter lower Earth Parking Orbit. Even the mighty F-1 engines were reorificed to provide more "bang-for-the-buck". The additional 4,000 pounds of payload capacity would be spoken for right down to the ounce.

The most significant addition of equipment was the first lunar car. For the first time a crew of explorers were to drive a vehicle on another world. The Lunar Rover was NASA's version of the stalwart military Jeep.

In the early 1940's the United States Army asked contractors to submit proposals for an all-terrain vehicle which could meet a series of stringent requirements. It had to be tough, lightweight, high-powered and capable of enduring harsh environments. The car had to be delivered for trials within 49 days. With only hours to spare, on 23rd September 1940, the Bantam motor company delivered the *Bantam* **LRV** (Light Reconnaissance Vehicle)—the first Jeep, and a legend was born. The *Bantam* had a weight limit of 1275 lbs and ran on a 45 horse-power engine. Its payload was 600 lbs and it had a low crawling speed of 5 kph. It was able to navigate steep inclines and it used four wheel drive. It was also capable of low-speed cooling. As remarkable as the original Jeep was, it barely holds a candle to the *Boeing* **LRV** (Lunar Roving Vehicle).

The *Boeing* LRV was an all-terrain vehicle which could be folded in half. It was delivered to NASA on March 14th 1971, with two weeks to spare, and 17 months after the signing of the contract. It had a "curb-weight" of 462 lbs (on the Earth) and yet it was capable of carrying 1080 lbs of cargo. It used four electric motors that delivered a total of one horse-power to its four-wheel drive system and it had a top speed of 14 kph. It was able to maintain its balance in one-sixth gravity on slopes up to 45°, while its thermal control had to contend with temperatures ranging from -200 to +400 degrees F°. Its wheels could withstand driving over solid rock without the benefit of inflatable tires. It had double redundancy on most of its major components, including steering. It was powered by two electric batteries while carrying a portable television station with a range of at least a quarter of a million miles. It also had an onboard navigation system that would tell the driver where he was, where he was going and where he had been. Slightly more than thirty years had passed but I'll wager Karl Probst, the designer of the *Bantam*, would have been proud and amazed. Even if Boeing did have a bigger budget...

After thirty five years the sheer audacity of the full-up "J" missions has been forgotten by most people. Apollo 15 was the first and arguably the most seductive of these flights in part due to the astonishing landscape of Hadley-Apennine. David Scott and Jim Irwin would board their fully-loaded Lunar Module and fly it straight down between mountains almost three miles high and land amidst one of the richest geologic treasure troves ever encountered by man. Not to be outdone, Command Module Pilot Alfred Worden would run through a daunting schedule of science for three days from lunar orbit. The so-called "SIM" bay aboard the *Endeavour* housed a battery of scientific equipment which would photograph, catalog, scan and probe the lunar surface.

The Apollo 15 flight plan would be similar to that of Apollo 14 wherein the command and service modules would be brought down to a very low altitude before undocking with the LM. On Apollo 15 that would be only nine and a half miles from the surface. Just over 100 hours into the mission the two vehicles separated and four hours later Commander Scott began the short descent into the Hadley mountains. Twelve minutes and twenty seconds later the *Falcon* lunar module settled into one of the most startlingly beautiful valleys ever seen by human eyes. Hadley had been chosen because it boasted a dazzling array of strange geological formations not least of which was the enormous rift known as Hadley Rille, a sinuous snaking gorge that stretched like an old river across over 100 kilometers of lunar terrain. It is almost half a kilometer deep (1,300 feet) and three times as wide. Reaching Hadley presented a host of new challenges due to its high latitude (26°N).

However, once on the surface, Scott and Irwin exemplified how a well trained expert can do a great job, even in the most alien of situations. Spread over two and a half days the two astronauts spent more than 18 hours on the lunar surface and, thanks to the remarkable LRV, traveled nearly 28 kilometers. Their first full EVA was preceded by a short 33 minute excursion through the top hatch of the LM by Scott. This was called an SEVA (standup extravehicular activity) and was done to try and get a good look at the surrounding topography from the highest point available. After a rest they exited the LM, deployed the rover and took off south-westwards towards the Rille. They investigated this unique location before they returned and deployed another ALSEP. Sixteen hours later they were off again, this time to study a dense cluster of craters to the south, called predictably *South Cluster,* before heading onto the base of the enormous, 10,000 foot tall, Mount Hadley Delta. It was here at a location dubbed Station 7 that Scott found what the geologists were looking for, an ancient kind of rock called anorthosite. Later nicknamed the *Genesis* rock it proved to be one of the oldest rocks ever found, dating back over four billion years and it bolstered many geologists' models of the formation of the moon. Scott and Irwin took another fourteen hour break before taking one more drive. Cancelling plans to go three kilometres north to another crater cluster, because they were running behind schedule, instead they took another trip to the Rille about 1800 m to the west of the LM. They stopped off at the ALSEP and retrieved experiment core samples before finally retiring to the spacecraft in preparation for launch just over three and a half hours later.

After accomplishing everything in the flight-plan, and more, *Falcon* lifted off to a perfect rendezvous with *Endeavour,* an event that was broadcast for the first time on live television thanks to the batteries aboard the LRV. On the return voyage CM pilot Al Worden would become the first human to perform a space walk in deep space to retrieve film cartridges, while presumably becoming the first person ever to see an unobscured view of the Earth and the Moon in their entirety, simply by turning his head.

The old expression in the movie business is that without drama there is no audience. The element of risk is what draws people to watch. With that axiom in mind it seems incredible that by the time Apollo 16 left for the moon in April 1972 most of the world had stopped watching. The great bulk of the American audience were completely wrapped up in what was happening in Viet Nam and showed little or no interest in what was happening in the lunar highlands.

What was transpiring a quarter of a million miles away was as remarkable as any of the great voyages of exploration in human history. Just as Livingston or Shackleton had taken enormous risks to further man's knowledge of his own world, John Young and Charlie Duke were rolling the dice to find out more about our nearest neighbor. The American public seemed to have resigned themselves to the notion that nothing extraordinary was going to happen at Descartes and so they simply stopped watching. The novelty of the lunar rover had worn off and there was no high drama, such as that epitomized by the dilemma of Apollo 13. Apollo 16 was a victim of its own success. John Young, Charles Duke and Ken Mattingly flew an almost flawless mission but most people didn't notice.

Apollo 16 was not about the Moon—it was about understanding the Earth. John Young and Charlie Duke were given the task of landing their lunar module in the highland area of Descartes. It was approximately 7,400 feet higher elevation than Tranquillity Base. The object? To bring back samples of lunar bedrock or evidence of lunar volcanism.

Geology is certainly not the most glamorous science. It involves clambering around in the dirt for hours and hours, digging holes, cracking rocks with a hammer and drilling core samples. For many of the general public this represents a situation beyond the threshold of boredom and they certainly can't be bothered to watch it on television. The fact that these two explorers were conducting geology on another world was lost in amongst the perfect color television pictures—if not for the pitch black sky Duke and Young could just as easily have been somewhere in the deserts of the United States.

Young and Duke went out the door of their lunar home for three days straight. They were a quarter of a million miles from any kind of support and they were absolutely plastered in lunar dirt. It was so bad that each day they wondered if they might not be able to seal their gloves and helmets. The regolith acted like an abrasive and everything it came into contact with was scratched. As if that wasn't bad enough the orange juice dispensers were leaking inside their helmets, making the work environment sticky and unpleasant.

Ignoring these risks and discomforts, Young and Duke plunged into their work with enthusiasm. 119 hours into the mission they depressurized their lunar module and ventured outside where they were almost immediately beset with bad luck. Young accidentally got his foot caught in a power cable and inadvertently ripped it from its connectors, effectively killing one experiment. Then the rear steering on the rover wouldn't work but inexplicably started working forty minutes later, just in time for the first traverse.

The landing site at Descartes was chosen because it appeared to have been caused by volcanic activity. Confirmation of such volcanism would reveal important information about the origins of the Moon—and the Earth. The absence of such volcanic activity would also change many of the conventional theories back here on Earth. As it turned out Descartes was not volcanic but it was very old, nearly 4.5 billion years. Some of the rocks even predated the famous anorthosite brought back by Apollo 15. Theories of lunar volcanoes which hailed back to the days of Galileo were finally put to rest by Apollo 16. Young and Duke also set up an ultra-violet observatory on the Moon to study the effects of solar radiation in an environment unimpeded by the Earth's atmosphere. As on previous missions they established a sophisticated science station which would relay information about the lunar environment back to Earth for years to come. The whole time they were on the surface, Thomas "Ken" Mattingly, aboard the orbiting CSM, reeled in copious amounts of data from orbit using the onboard Scientific Instrument Module.

Altogether Young and Duke were on the lunar surface for 73 hours and they spent over 20 hours of that time outside. They traversed over 27 kilometers in their rover, ultimately returning to the Earth with 97 kilograms of moon rocks.

The stage was now set for the final lunar mission of the 20th century. As has been mentioned earlier the subsequent missions had been cancelled but this left one awkward dilemma. Apollo 18 was to have sent a fully trained scientist to the moon, it was to have been the big payoff for the thousands of hours of effort put in by the geology community, but now, that flight had been cancelled. Recognizing that this was to be the final voyage, NASA decided to change the crew. Lunar Module pilot Joe Engle who had been as eager and as qualified as any of his peers and had gone no further in the Apollo program than flying a lunar mission in a pressure chamber in Houston was to be removed from the flight roster and replaced at the last minute by Apollo 18 LM pilot Harrison "Jack" Schmitt. No one could argue that it made sense but in retrospect it must have been a bitter blow for Engle.

Once this disagreeable situation had been resolved Saturn/Apollo 512 awaited its moment on the launch pad. It was December 6th 1972 and there was a delay. Tens of thousands of people had come to watch this final launch and this time it was to be even more spectacular. The launch would take place at night. No one knew what a Saturn V would look like at night but when the moment came the voyeurs gasped in astonishment. Although it was 12.33 a.m. on December 7th it was as though daytime had come to Florida. Entire cruise ships had been chartered to watch the beginning of the end. Just off-shore one such ship housed no lesser luminaries than Robert Heinlein, Isaac Asimov and Norman Mailer, this was a show not to be missed.

Just over two hours and forty minutes late Apollo 17 thundered into the night sky. This delay still leads to some confusion in the documentation, the official Mission Report times seem to rarely be consistent. Some show the expected times and some show the actual. On this mission Commander Eugene Cernan had the unprecedented ability to manually fly the giant booster during this most dangerous part of the mission, but all went well and soon the last lunar Apollo was in an orbit just 90 miles above the Earth. Just as before, a little over three hours into the mission, the super-reliable J-2 engine, attached to the third stage, re-ignited and pushed the 50,000 pounds of hardware away from the Earth and on-course for the moon. An hour later the lunar module was successfully extracted and the third stage was sent on its way to a collision with the moon.

The accuracy of the launch was such that only one of four planned course corrections proved to be necessary. 89 hours after launch Apollo 17 settled into lunar orbit, 26 minutes later the third stage of the Saturn V impacted the moon and set the whole planet ringing like a bell, setting off the seismometers at all of the previous ALSEP sites and making geologists and planetary scientists around the world rub their hands with glee.

Four hours later Ron Evans, the command module pilot, fired the main engine and began the descent towards the surface. Once the two craft were in an elliptical orbit with the low point only 50,000 feet above the surface Cernan and Schmitt prepared for the final descent to the valley of Taurus-Littrow.

Once again, the LM engine burn took just over 12 minutes before the delicate lander arrived right on target, a little under five days after leaving the Earth. Four hours later Cernan exited the LM and Schmitt came close behind.

For the next three days (actually 75 hours) the test pilot and the scientist traveled and worked together over the entire expanse of the starkly beautiful valley. On their first EVA they unloaded their rover, deployed the flag and the ALSEP experiments package, took some snapshots and then after being outside for nearly five hours headed south for a thirteen minute drive to *Steno* crater. After an hour and a quarter of exploring they returned to the LM, having deployed some more experiments and taken some panoramic photographs. Seventeen hours later they were again back in the rover and heading due west towards what would become known as *Station 2*. It was located in an area called *Nansen* and it took them an hour and ten minutes to get there (allowing for a few stops along the way.) This was the furthest afield that any Apollo astronaut would ever venture, over eight kilometers (about five miles) from the LM. Next they moved north-east to the edge of a crater called *Lara*, only about two kilometers distant, but again making stops along the way to carefully choose samples, this drive took 41 minutes. At this point the two explorers had been outside for just under four hours and they were still literally miles from home. Next they headed north-east to *Shorty* crater where they spent just over half an hour before turning south-east and back towards the largest crater in the valley, known as *Camelot*.

Time was now running out and they were still more than a kilometer from the LM, so they only spent a short half hour at *Camelot* before turning east back to the ALSEP site. They had been outside for nearly eight hours and they were exhausted but jubilant. Schmitt's expertise proved invaluable but Cernan also demonstrated that his intense training was far from wasted. Later that day (it was technically all one day since a day on the moon lasts over three hundred hours) the rover was loaded up for the last time. They had rested for over fifteen hours and on this last jaunt they would head due north towards the imposing base of a mountain range known as the North Massif. Station 6 was almost two miles away at the foot of the mountain and allowing for two short stops on the way they averaged about four miles an hour.

Again they collected samples and took stupendous panoramic photographs of the whole valley laid out below them, before heading east, across the mouth of a narrow gorge called the *Wessex Cleft,* to the base of the *Sculptured Hills.* Again they took gravimetric readings, shot pictures and gathered samples, at this point they had been outside for just over three hours. Heading south west they cruised downhill to *Van Serg* crater. The two kilometer ride took 17 minutes which translates to about 3.5 kph. This final stop occupied them for almost an hour. They deployed one of several explosive charges, dug a trench for sampling, and took more photos before heading for home. The last lunar rover had run them around the valley for nearly six hours on this final excursion and they now positioned it in such a way so that its television camera could watch them lift off. This time they would be given an opportunity to rest before leaving. An eighteen hour break preceded the final departure of the last humans to walk on the moon.

Cernan, Schmitt and Evans had been in space for nearly eight days when the lunar module *Challenger* finally soared up toward its rendezvous with the command module *America*. It was carrying 250 pounds of rocks and two very tired astronauts. The journey home was uneventful, almost five days of coasting downhill to a perfect splashdown in the Pacific ocean after 12½ days in space.

Apollo was not totally finished. There were still four more manned missions, three to the giant space station Skylab and one on a joint mission with the Soviet Union. All four flights were enormously successful and important for different reasons but none would capture the world's collective imagination like the moon landings.

And so ended the first explorations of the moon. Decades have now passed and scientists around the world are still studying the rocks and bouncing lasers off the mirrors left on the moon. Millions of under-educated people think the whole program was faked, for the most part they seem to be the same people who are prepared to believe in anything that science says *doesn't* exist. These sceptics are the product of an era where science and engineering are disregarded in schools and have become side-shows to the inanities of "reality" television and where fanaticism in politics once again threatens the truth, the way it did for Galileo and Aristarchus. The era of Apollo may have been concurrent with the era of the Cold war and the hell on earth that was the Viet Nam war, but those sordid episodes will be long forgotten while Apollo will inevitably endure.

At the time of this writing NASA and some industrious members of the private sector are drawing up plans to return to the moon. Some of the drawings look suspiciously like Apollo but all of these ideas are burdened by the absence of a heavy lift launch vehicle like the Saturn V. The dismantling of the Saturn infrastructure poses the greatest challenge to the implementation of these plans. Once Apollo had won the space race NASA returned to the 1950's paradigm for space exploration—build a shuttle to service a space station which could then be used as a staging post to launch manned missions to the moon and beyond. Although the goal of a reusable vehicle was a worthy one, what seems to have been overlooked was the magical decade of Saturn/Apollo that made most of those schemes irrelevant and in hindsight seems to have superceded them in every way.

The twelve men who walked on our sister planet, the most prestigious and exclusive club of humans in history, are now reduced to nine. Jim Irwin, Al Shepard and Pete Conrad all passed away before their time, but they left the kind of legacy that fill the dreams of kings and presidents. All in the short space of three years these twelve men, along with dozens of their peers and the hundreds of thousands of people around the world who had contributed to Apollo, proved that humankind are a species with potential. Fragile as we are, they proved we are tough enough, smart enough, and brave enough to break free of our tiny planetary cradle and snatch greatness from the heavens.

APOLLO/SATURN FLIGHTS

Mission	Launch Date	Launch Vehicle	Payload	Description
APOLLO 11	7/16/69	SA-506	CM-107 SM-107 LM-5 SLA-14	First manned lunar landing mission. Lunar surface stay time 21.6 hours. Mission duration 8 days 3 hours. EASEP
APOLLO 12	11/14/69	SA-507	CM-108 SM-108 LM-6 SLA-15 ALSEP I	2nd manned lunar landing mission. Demonstration of point landing capability. Deployment of ALSEP I. Surveyor III investigation. Lunar surface stay time 31.5 hours. Two dual EVA's (15.5 man hours). 89 hours in lunar orbit (45 orbits) Mission duration 10 days 4.6 hours.
APOLLO 13	4/11/70	SA-508	CM-109 SM-109 LM-7 SLA-16 ALSEP III	Planned 3rd lunar landing. Mission aborted at approximately 56 hOUrs due to loss of SM cryogenic oxygen and consequent loss of capability to generate electrical power and water.
APOLLO 14	1/31/71	SA-509	CM-110 SM-110 LM-8 SLA-17 ALSEP 14	3rd successful lunar landing mission. Landing at Fra Mauro site. Deployment of ALSEP. Extensive geology traverse. Lunar stay time 34.5 hours. Two dual EVA's of 4 hr. 49 min. and 4 hr. 28 min. Mission duration 9 days 2 min.
APOLLO 15	7/26/71	SA-510	CM-112 SM-112 LM-10 SLA-19 ALSEP 15 LRV-1	4th successful lunar landing mission. Exploration at Hadley-Apennine site. Extensive geology traverses with first lunar roving vehicle (27.9 km). Deployment of ALSEP. Lunar stay time 66.9 hours. Three dual EVA's totaling 18.6 hrs. 145.3 hrs in lunar orbit (74 orbits). Mission duration 12 days 7.2 hrs.
APOLLO 16	4/16/72	SA-511	CM-113 SM-113 LM-11 SLA-20 ALSEP 16	5th successful lunar landing mission. Exploration at Descartes site. Deployment of ALSEP and other experiments. Three extensive geology LRV traverses. Three dual EVA's totaling 20.3 hours. Lunar stay time 71 hours. 2nd use of scientific instrument module for orbital science. 126.1 hours in lunar orbit (64 orbits). Mission duration 11 days 1.8 hours.
APOLLO 17	12/7/72	SA-512	CM-114 SM-114 LM-12 SLA-21 ALSEP-17 LRV-3	6th lunar landing mission. Exploration at Taurus-Littrow site. Deployment of ALSEP and other experiments. Three extensive geology traverses on LRV. Three dual EVA's totaling 88.4 hrs. Lunar stay time 75 hrs. 3rd use of scientific instrument module for orbital science. 75 lunar orbits. Mission duration 12 days 14 hours.

APOLLO 11 (AS-506) FLIGHT SUMMARY

GENERAL

Spacecraft: CM-107, SM-107, LM-5
Launch Vehicle: SA-506
Launch Complex: 39A
Flight Crew:
Commander (CDR) Neil A. Armstrong
Command Module Pilot (CMP) Michael Collins
Lunar Module Pilot (LMP) Edwin E. Aldrin, Jr.
Launch Time: 9:32 a.m. EDT, July 16, 1969
Flight Azimuth: 72°
Earth Orbit: 102.9 x 103.7 NM
Lunar Orbits and Events:
Initial Apocynthion/ Pericynthion (LOI-1): 168.6 x 61.2 NM
Circularized Apocynthion/ Pericynthion (LOI-2): 65.7 x 53.8 NM LM
Descent Orbit: 57.2 x 8.5 NM
Landing Site Coordinates: 0.647°N. latitude, 23.505° E. longitude (Tranquility Base)
Lunar Landing Time: 4:17:40 p.m. EDT, July 20, 1969
First Step on Lunar Surface: 10:56:19 p.m. EDT, July 20, 1969
LM Liftoff from Lunar Surface: 1:54:00 p.m. EDT, July 21, 1969
Lunar Insertion Orbit: 45.2 x 9.0 NM
Final LM/CSM Separation Orbit: 62.6 x 54.8 NM
Mission Duration: 195 hrs 18 mins 35 secs
Time of Landing: 12:50:35 p.m. EDT, July 24, 1969

APOLLO 11

SPACE VEHICLE AND PRE-LAUNCH DATA

Spacecraft delivered to KSC:
Command/service module: January 1969
Lunar module: January 1969
Launch vehicle delivered to KSC:
First stage (S-IC): February 1969
Second stage (S-II): February 1969
Third stage (S-IVB): January 1969
Instrument unit (IU): February 1969
Space vehicle weight at liftoff: 6,398,325 lb.
Weight placed in earth orbit: 297,848 lb.
Weight placed in lunar orbit: 72,038 lb.

MISSION PRIMARY OBJECTIVE

(Accomplished) Perform a manned lunar landing and return.

DETAILED OBJECTIVES AND EXPERIMENTS

(All Accomplished unless noted)
1. Collect a contingency sample.
2. Egress from the LM to the lunar surface, perform lunar surface EVA operations, and ingress into the LM from the lunar surface.
3. Perform lunar surface operations with the EMU.
4. Obtain data on effects of DPS and RCS plume impingement on the LM and obtain data on the performance of the LM landing gear and descent engine skirt after touchdown.

5. Obtain data on the lunar surface characteristics from the effects of the LM landing.

6. Collect lunar Bulk Samples.

7. Determine the position of the LM on the lunar surface.

8. Obtain data on the effects of illumination and contrast conditions on crew visual perception.

9. Demonstrate procedures and hardware used to prevent back contamination of the earth's biosphere.

10. Deploy the Early Apollo Scientific Experiments Package (EASEP) which included the following:

a. S-031, Passive Seismic Experiment.

b. S-078, Laser Ranging Retro-Reflector.

11. Deploy and retrieve the Solar Wind Composition Experiment, S-080.

12. Perform Cosmic Ray Detector Experiment (helmet portion), S-151.

13. Perform Lunar Field Geology, S-059. (Partially Accomplished)

14. Obtain television coverage during the lunar stay period.

15. Obtain photographic coverage during the lunar stay period.

UNUSUAL FEATURES OF THE MISSION

1. First manned lunar landing and return.

2. First lunar surface EVA.

3. First seismometer deployed on moon.

4. First laser reflector deployed on moon.

5. First solar wind experiment deployed on moon.

6. First lunar soil samples brought to earth.

7. Largest payload yet placed in lunar orbit.

8. First lunar module test in total operational environment.

9. Acquisition of numerous visual observations, photographs, and television of scientific and engineering significance.

10. First operational use of the mobile quarantine facility (MQF) and the lunar receiving laboratory (LRL).

Significant spacecraft differences from Apollo 10:

Command/Service Module

* The blanket type insulation was removed from the forward hatch.

Lunar Module

* A VHF antenna was added for extravehicular activity (EVA) coverage.

* A liquid cooling garment (LCG) heat removal subsystem was added.

* The ascent engine was replaced with a lighter weight engine.

* The base heat shield on the descent stage was modified by the removal of H-film.

* Reaction control system (RCS) plume deflectors were added for each of the lower four RCS thrusters.

* The landing gear thermal protection was increased.

* The descent propulsion system (DPS) engine gimbal drive actuators were modified by the removal of the polarizer and armature and by the installation of new brake material.

* An erectable S-band antenna was carried on the descent stage.

* The Early Apollo Scientific Experiments Package (EASEP) was carried in the descent stage.

Significant launch vehicle changes from Apollo 10:

S-IC Stage

* Research and development instrumentation was deleted. Only operational instrumentation was retained.

* The accumulator bottles were

deleted from the pneumatic control system.

S-II Stage
* The insulation on the engine start tank was improved..
* Cork insulation was added over the spray foam in hot-spot ramp areas.
* An automatic checkout system was incorporated for the inflight helium injection system.

S-IVB Stage
* Additional Instrumentation for the O2H2 burner was installed and wired to the IU.

RECOVERY DATA

Recovery Area: Mid-Pacific Ocean
Landing Coordinates: 13°15.25' N., 169°09.4' W. (Stable II)
Recovery Ship: USS Hornet Crew
Recovery Time: 1:57 p.m. EDT, July 24, 1969
Spacecraft Recovery Time: 3:57 p.m. EDT, July 24, 1969

REMARKS

The first manned lunar landing mission was conducted as planned and all primary mission objectives were successfully accomplished. The accuracy of injection maneuvers and trajectories was such that only one midcourse correction during translunar coast and one during transearth coast were required. As a result of Apollo 10 lunar orbit experience, the LOI-2 burn was biased to achieve a slightly eccentric orbit (65.7 x 53.8 NM). It was anticipated that this would compensate for variations in lunar gravity effect and that the CSM orbit would become circular by the time of LM rendezvous. Subsequent measurements showed that this effect did not occur as rapidly as expected and that the CSM orbit did not become circular. The LM powered descent initiation maneuver was performed on time at pericynthion on the descent orbit; however, this position was about 4 NM downrange from the planned point apparently due to an accumulation of uncoupled attitude maneuvers during the last two revolutions prior to PDI. This resulted in the landing point being shifted downrange about 4 NM. During the final approach phase, the crew noted that the LM was headed for the general area of a large, rugged crater, filled with boulders of 5 to 10 feet in diameter. The CDR took manual attitude control and translated the LM to a landing point approximately 1000 feet farther downrange. The crew adapted quickly to the lunar environment and conducted the lunar surface activities as planned, including the collection of two lunar core samples and a considerable amount of discretely selected surface material. The LMP had to exert a considerable force to drive the core tubes an estimated 6 to 8 inches deep. The crew spent a total of 5 manhours of EVA on the lunar surface. The total lunar stay time was 21 hours 36 minutes. Approximately 46 pounds of lunar samples were returned to earth. All launch vehicle systems performed satisfactorily throughout their expected lifetimes and all spacecraft systems continued to function satisfactorily throughout the mission. No major anomalies occurred. New biological isolation procedures and post- recovery operations were executed successfully. Flight crew performance was outstanding and all three crew members remained in excellent health.

APOLLO 12 (AS-507) FLIGHT SUMMARY

GENERAL
Spacecraft: CM-108, SM-108, LM-6
Launch Vehicle: SA-507
Launch Complex: 39A
Flight Crew:
Commander (CDR) Charles Conrad
Command Module Pilot (CMP) Richard F. Gordon
Lunar Module Pilot (LMP) Alan L. Bean
Launch Time: 11:22 a.m. EST, November 14, 1969
Launch Azimuth: 72°
Earth Orbit: 102.5 x 99.9 NM
Lunar Orbit and Events:
Initial Apocynthion/ Pericynthion (LOI-1): 168.8 x 62.6 NM
Circularized Apocynthion/ Pericynthion (LOI- 2): 66.1 x 54.3 NM
LM Descent Orbit: 60.6 x 8.1 NM
Landing Site Coordinates: 3.036° S. latitude, 23.418°W. longitude
Lunar Landing Time: 01:54:35 a.m. EST, November 19, 1969
LM Liftoff from Lunar Surface: 09:25:47 a.m. EST, Nov 20, 1969
Lunar Insertion Orbit: 46.3 x 8.8 NM
Ascent Stage Impact on Lunar Surface: 5:17:16 p.m. EST, Nov 20, 1969
Ascent Stage Impact Coordinates: 3.95°S. lat, 21.17°W. long
Ascent Stage Impact Velocity: 5502 fps

APOLLO 12

Ascent Stage Impact Weight: 5254 pounds
Mission Duration: 244 hrs 36 mins 24 secs
Time of Landing: 3:58 p.m. EST, November 24, 1969

SPACE VEHICLE AND PRE-LAUNCH DATA

Spacecraft delivered to KSC: Command/service module: March 1969
Lunar module: March 1969
Launch vehicle delivered to KSC:
First stage (S-IC) : May 1969
Second stage (S-II) : May 1969
Third stage (S-IVB) : May 1969
Instrument unit (IU): May 1969
Space vehicle weight at Liftoff: 6,484,780 lb.
Weight placed in earth orbit: 300,056 lb.
Weight placed in lunar orbit: 72,212 lb.

MISSION PRIMARY OBJECTIVES
(All Primary Objectives Accomplished)
1. Perform selenological inspection, survey, and sampling in a mare area.
2. Deploy and activate the Apollo Lunar Surface Experiments Package (ALSEP).
3. Develop techniques for a point landing capability.

4. Develop man's capability to work in the lunar environment.

5. Obtain photographs of candidate exploration sites.

DETAILED OBJECTIVES AND EXPERIMENTS

PRINCIPAL

(All Principal Detailed Objectives Accomplished)

1. Collect a contingency sample.

2. Perform lunar surface EVA operations.

3. Deploy ALSEP I which included the following:

  a. S-031, Passive Seismic Experiment.

  b. S-034, Lunar Surface Magnetometer Experiment.

  c. S-035, Solar Wind Spectrometer Experiment.

  d. S-036, Suprathermal Ion Detector Experiment.

  e. S-058, Cold Cathode Ionization Gauge Experiment.

  f. M-515, Lunar Dust Detector.

4. Collect selected samples.

5. Recharge the portable life support systems.

6. Perform Lunar Field Geology, S-059.

7. Obtain photographic coverage of candidate exploration sites.

8. Obtain data on the lunar surface characteristics from the effects of the LM landing.

9. Obtain data on the effects of illumination and contrast conditions on Crew visual perception.

10. Determine the position of the LM on the lunar surface.

11. Perform selenodetic reference point update.

12. Deploy and retrieve the Solar Wind Composition Experiment, S-080.

13. Perform Lunar Multispectral Photography Experiment, S-158.

SECONDARY

1. Investigate and obtain samples for earth return from the Surveyor III spacecraft. (Accomplished)

2. Obtain photographic coverage during the lunar stay period. (Accomplished)

3. Obtain television coverage during the lunar stay period. (Partially Accomplished)

UNUSUAL FEATURES OF THE MISSION

1. First use of the S-IVB stage to perform an evasive maneuver.

2. First use of a hybrid trajectory.

3. Largest payload yet placed in lunar orbit.

4. First demonstration of a point landing capability.

5. First use of two lunar surface EVA periods (about 4 hours each).

6. First ALSEP deployed on the moon.

7. First deployment of the erectable S-band antenna.

8. First recharge of the portable life support system.

9. First documented samples returned to earth.

10. First use of geologists to plan a lunar surface traverse in real time.

11. First double core-tube sample taken.

12. First return of samples from a prior lunar landed vehicle (Surveyor III).

13. Longest distance yet traversed on the lunar surface.

14. First multispectral photography from lunar orbit.

15. Longest lunar surface stay to date.

16. Longest lunar mission to date.

17. Largest payload yet returned from the lunar surface.

Significant spacecraft differences from Apollo 11:

Command/Service Module

* Experiment S-158 was incorporated and the side hatch window pane was changed for Lunar multispectral photography.
* The reaction control system (RCS) engine arc was suppressed.
* An inertial measurement unit (IMU) Power switch guard was added.
* Stowage was modified to provide for return of Surveyor III samples and increased lunar surface samples.

Lunar Module

* The display and keyboard assembly (DSKY) table and support were modified to enhance actuation and release from the stowed to the operating position.
* The ascent stage propellant tanks were redesigned to an all-welded configuration.
* Stowable hammocks were added for increased crew sleeping comfort.
* The bacteria filter was deleted from the forward hatch valve.
* Stowage was modified to provide for return of Surveyor III samples and increased lunar surface samples.
* Landing gear and plume deflector thermal insulation was reduced.
* Extravehicular activity (EVA) equipment stowage was modified.
* Apollo Lunar Surface Experiments Package (ALSEP) was installed to replace the Early Apollo Scientific Experiments Package (EASEP).

Significant launch vehicle changes from Apollo 11:

S-IVB Stage

* The telemetry system for the S-IVB stage was changes by adding one SSB/FM link to provide increased acoustic and vibration measurements.

RECOVERY DATA

Recovery Area:
Mid-Pacific Ocean
Landing Coordinates: 15°47'S., 165°11'W. (Stable II)
Recovery Ship: USS Hornet
Crew Recovery Time: 4:58 p.m. EST, November 24, 1969
Spacecraft Recovery Time: 5:49 p.m. EST, November 24, 1969

REMARKS

Launch vehicle performance was satisfactory throughout its expected lifetime except for the S-IVB slingshot maneuver. The spacecraft systems functioned satisfactorily during the entire mission except for the perturbations caused by an electrical anomaly which occurred shortly after liftoff. Communications were very good except for occasional problems with the high gain antenna (HGA). The spacecraft and launch vehicle were involved in two electrical potential discharges during the first minute of the flight. The first, at 36.5 seconds after liftoff, was from the clouds to earth through the vehicle and was visible to launch site observers. The second occurred at 52 seconds with the vehicle in the clouds. The discharge at 36.5 seconds disconnected the fuel cells from the spacecraft buses and damaged nine instrumentation measurements. The discharge at 52 seconds caused tumbling of the spacecraft inertial platform. Both discharges caused a temporary interruption of spacecraft communications. Many other effects were noted on instrumentation data from the launch vehicle, which apparently sustained no permanent malfunctions from the discharges. The S-IVB slingshot maneuver was initiated on schedule but, due to IU state vector errors, the slingshot maneuver did not achieve the desired heliocentric orbit but rather a highly eccentric geocentric orbit. Lunar orbit insertion (LOI) was performed in two separate maneuvers, LOI-1 and LOI-2, using the Service propulsion system (SPS). The LOI maneuver resulted in a CSM/LM position some 4 to 5 NM north of the expected ground track prior to descent orbit insertion (DOI). This cross-range error was known prior to DOI and was corrected during the powered descent maneuver. The guidance computer was updated during powered descent to compensate for indications that the trajectory was coming in 4200 feet short of the target point. The initial cross-range distance was continuously reduced throughout the braking phase. At entry into the approach phase the Spacecraft's trajectory was very close to nominal. Redesignations were incorporated during the approach phase. The crew took over manual control at about 370 feet, passed over the right side of the target crater, then flew to the left for landing. The commander reported extensive dust obscuring his view during final descent. The actual landing point is determined to be about 600 feet from the Surveyor III spacecraft. The ascent stage deorbit retrograde burn was initiated and burned slightly longer than planned. This resulted in lunar impact about 36 NM short of the target point. Impact occurred about 39 NM southeast of Surveyor III. On several occasions during the mission, communications with the CSM experienced some degradation due to inability of the HGA to hold lock. Two special HGA tests were conducted during the transearth coast to attempt to identify the cause of the anomaly. Results indicate that the problem appears to be associated with the dynamic thermal operation of the antenna, probably in the microwave circuitry in the narrow beam mode. The Apollo 12 crew performance was outstanding throughout the mission. All scheduled lunar surface scientific activities were performed as planned within the allotted time periods. During the first EVA the ALSEP experiments were deployed and began transmitting scientific data. Real-time planning for the geological traverse of the second EVA was accomplished jointly by the crew and earth-based scientists. All planned Surveyor activities were performed and, in addition, retrieval of the Surveyor scoop containing a surface sample was accomplished.

Approximately 75 pounds of samples were collected during the two 2-man EVA's which totaled 7 hr. 45 min. The traverse distance was approximately 2 km.

APOLLO 13 (AS-508) FLIGHT SUMMARY

GENERAL

Spacecraft: CM-109, SM-109, LM-7
Launch Vehicle: SA-508
Launch Complex: 39A
Flight Crew:
Commander (CDR) James A. Lovell, Jr.
Command Module Pilot (CMP) John L. Swigert, Jr.
Lunar Module Pilot (LMP) Fred W. Haise, Jr.
Launch Time: 2:13 p.m. EST, April 11, 1970
Launch Azimuth: 72°
Earth Orbit: 100.2 x 98.0 NM
Closest Approach to the Lunar Surface: 142.8 NM
S-IVB/IU Lunar Impact Time: 8:09:40 p.m. EST, April 14, 1970
Velocity of Impact: 8465 fps
Angle of Impact: Appx. 80° to the horizontal
Lunar Location: 2.4°S., 27.9°W.
Energy Equivalent: 11.5 tons of TNT
Mission Duration: 142 hrs 54 mins 41 secs
Time of Landing: 1:07:41 p.m. EST, April 17, 1970

SPACE VEHICLE AND PRE-LAUNCH DATA

Spacecraft delivered to KSC:
Command/service module: June 1969
Lunar module: June 1969

APOLLO 13

Launch vehicle delivered to KSC:
First stage (S-IC); June 1969
Second stage (S-II): June 1969
Third stage (S-IVB) : June 1969
Instrument unit (IU): July 1969
Space vehicle weight at liftoff: 6,421,259 lb.
Weight placed in earth orbit: 296,463 lb.

MISSION PRIMARY OBJECTIVES

(None Accomplished)
1. Perform selenological inspection, survey, and sampling of materials in a preselected region of the Fra Mauro Formation.
2. Deploy and activate an Apollo Lunar Surface Experiments Package (ALSEP).
3. Develop man's capability to work in the lunar environment.
4. Obtain photographs of candidate exploration sites.

DETAILED OBJECTIVES LAUNCH VEHICLE

(Secondary Objectives - Both Accomplished)
1. Impact of the expended S-IVB/IU on the lunar surface within 350 km of the targeted impact point of 3°S., 30°W. under nominal flight profile conditions to excite ALSEP I.
2. Post-flight determination of actual S-IVB/IU point of impact within 5

km, and time of impact within 1 second.

SPACECRAFT AND LUNAR SURFACE

(None Accomplished)
1. Contingency Sample Collection.
2. Deployment of the Apollo Lunar Surface Experiments Package (ALSEP III), which included the following:
a. S-031 Lunar Passive Seismology.
b. S-037 Lunar Heat Flow.
c. S-038 Charged Particle Lunar Environment.
d. S-058 Cold Cathode Ionization Gauge.
e. M-515 Lunar Dust Detector.
3. Selected Sample Collection.
4. Lunar Field Geology (S-059).
5. Photographs of Candidate Exploration Sites.
6. Evaluation of Landing Accuracy Techniques.
7. Television Coverage.
8. EVA Communication System Performance.
9. Lunar Soil Mechanics.
10. Selenodetic Reference Point Update.
11. Lunar Surface Closeup Photography (S- 184).
12. Thermal Coating Degradation.
13. CSM Orbital Science Photography (Includes S- 182).
14. Transearth Lunar Photography.
15. Solar Wind Composition (S-080).
16. EMU Water Consumption Measurement.
17. Gegenschein From Lunar Orbit (S- 178).
18. Dim Light Photography.
19. CSM/LM S-Band Transponder Experiment (S-164).
20. Downlink Bistatic Radar Experiment (VHF Portion Only) (S-170)

UNUSUAL FEATURES OF THE MISSION

1. Use of backup CM pilot.
2. First aborted Apollo Mission.
3. First impact of the S-IVB/IU on the lunar surface.
4. First use of lunar module to provide emergency propulsion and life support after loss of service module systems.

Significant spacecraft differences from Apollo 12:

None significant to mission flown.

Significant launch vehicle differences from Apollo 12:

A fourth battery was added to the instrument unit to extend command communications systems tracking to assist S-IVB/IU lunar impact trajectory and corrections.

RECOVERY DATA

Recovery Area: Mid-Pacific Ocean
Landing Coordinates: 21°38'24" S., 165°21'42" W. (Stable I)
Recovery Ship: USS Iwo Jima
Crew Recovery Time: 1:53 p.m. EST, April 17, 1970
Spacecraft Recovery Time: 2:36 p.m. EST, April 17, 1970

REMARKS

The Apollo 13 Mission was planned as a lunar landing mission but was aborted en route to the moon after about 56 hours of flight due to loss of service module cryogenic oxygen and consequent loss of capability to generate electrical power, to provide oxygen and to produce water in the command/service module. Shortly after the anomaly, the command/service module was powered down and the remaining flight, except for entry, was made with the lunar module providing all necessary power, environmental control, guidance and propulsion. Launch vehicle performance was satisfactory through first stage (S-IC) boost and into second stage (S-II) boost until the S-II center engine shut down approximately 132 seconds early. Low frequency oscillations (14 to 16 hertz) were experienced on the S-II stage and resulted in the early shutdown. To compensate for the early center engine cutoff the remaining four engines burned approximately 34 seconds longer than initially planned. Resultant S-II stage cutoff velocity was 223 feet per second (fps) lower than planned. As a result, the S-IVB orbital insertion burn was approximately 9 seconds longer than predicted with cutoff velocity within about 1.2 fps of planned. Total launch vehicle burn time was approximately 44 seconds longer than predicted. At termination of the orbital insertion burn, a greater than 3-sigma probability of meeting translunar injection cutoff conditions existed with remaining S-IVB propellants. The TLI burn was nominal. The planned S-IVB evasive maneuver and the subsequent LOX dump and Auxiliary Propulsion System (APS) burn were accomplished as planned. The S-IVB/IU impacted the lunar surface at 77:56:40 GET (08:09:40 p.m. EST, April 14) at 2.4°S., 27.9°W. and the seismometer deployed during the Apollo 12 mission successfully detected the impact as a seismic signal 20 to 30 times larger and four times longer than that caused by the impact of the Apollo 12 LM ascent stage. The target impact point was 110 NM from the Seismometer. The actual impact point was approximately 35 NM from the target point and about 85 NM from the Seismometer. Spacecraft systems performance was nominal until the fans in cryogenic oxygen tank 2 were turned on at 55:53:18. About 2 seconds after energizing the fan circuit, a short was indicated in the current from fuel cell 3, which was supplying power to cryogenic oxygen tank 2 fans. Within several additional seconds, two other shorted conditions occurred. Electrical shorts in the fan circuit ignited the wire insulation, causing temperature and pressure increases within cryogenic oxygen tank 2. When the pressure reached the cryogenic oxygen tank 2 relief valve full-flow conditions of 1008 psia, the pressure began decreasing for about 9 seconds, at which time the relief valve probably reseated, causing the pressure to rise again momentarily. About 1/4 second later, a vibration disturbance was noted on the command module accelerometers. The next series of events occurred within a fraction of a second between the accelerometer disturbances and the data loss. A tank line burst, because of heat, in the vacuum jacket pressurizing the annulus and, in turn, caused the blow-out plug on the vacuum jacket to rupture. Some mechanism in bay 4 combined with the oxygen buildup in that bay to cause a rapid pressure rise which resulted in separation of the outer panel. The panel struck one of the dishes of the high-gain antenna. The panel Separation shock closed the fuel cell 1 and 3 oxygen reactant shut-off valves and several propellant and helium isolation valves in the reaction control system. Data were lost for about 1.8 seconds as the high-gain antenna switched from narrow beam to wide beam, because of the antenna being hit and damaged. Following recovery of the data, the vehicle had experienced a translation change of about 0.4 fps, primarily in a plane normal to bay 4. Cryogenic oxygen tank 2 pressure indication was at the lower limit readout value. The cryogenic oxygen tank 1 heaters were on, and the tank 1 pressure was decaying rapidly. Fuel cells 1 and 3 operated for about 2-1/2 minutes after the reactant valves closed. During this period, these fuel cells consumed the oxygen trapped in the plumbing, thereby reducing the pressure below minimum requirements and causing total loss of fuel cell current and voltage output from these two fuel cells. Fuel cell 2 was turned off about 2 hours later because of the loss of pressure from cryogenic oxygen tank 1. As a result of these occurrences, the CM was powered down and the LM was configured to supply the necessary power and other consumables. The CSM was powered down at approximately 58:40 GET. The surge tank and repressurization package were isolated with approximately 860 psi residual pressure (approximately 6.5 pounds of oxygen total). The primary water glycol system was left with radiators bypassed. The first maneuver following the incident was made with the descent propulsion system at approximately 61:30 GET and placed the spacecraft once again on a free-return trajectory, with the altitude of closest lunar approach raised to 143 miles. A maneuver that was performed with the descent engine 2 hours after passing pericynthion reduced the transearth transit time from about 76 hours to 64 hours and moved the earth landing point from the Indian Ocean to the South Pacific. Two Small transearth midcourse corrections were required prior to entry; the first occurring at about 105:18 GET using the descent propulsion system and the second at approximately 137:40 GET using the lunar module reaction control system. All LM systems performed satisfactorily in providing the necessary power and environmental control to the spacecraft. The requirement for lithium hydroxide to remove carbon dioxide from the spacecraft atmosphere was met by a combination of CM and LM cartridges since the LM cartridges alone would not satisfy the total requirement. The crewmen, with direction from Mission Control, built an adapter for the CM cartridges to accept the LM hoses. The service module was jettisoned at approximately 138 hours GET, and the crew observed and photographed the bay-4 area where the cryogenic tank anomaly had occurred. At this time, the crew remarked that the outer skin covering for bay-4 had been severely damaged, with a large portion missing. The lunar module was jettisoned about 1 hour before entry, which was performed nominally using the primary guidance and navigation system. The performance of the flight crew was excellent throughout the mission. Their ability to implement the new procedures developed and tested by the flight operations team was exceptional. Similarly, performance of ground based personnel, both NASA and contractor, in analyzing the problem, developing new procedures and in running the extensive tests and simulations required to verify them was outstanding.

APOLLO 14 (AS-509) FLIGHT SUMMARY

GENERAL
Spacecraft: CM-110, SM-110, LM-8
Launch Vehicle: SA-509
Launch Complex: 39A
Flight Crew:
Commander (CDR) Alan B. Shepard, Jr.
Command Module Pilot (CMP) Stuart A. Roosa
Lunar Module Pilot (LMP) Edgar D. Mitchell
Launch Time: 4:03 p.m. EST, January 31, 1971
Launch Azimuth: 75.56°
Earth Orbit:
100.2 x 99.2 NM
S-IVB/IU Lunar Impact Time: 1:01 a.m. EST, February 4, 1971
Velocity of Impact: 8,350 fps
Lunar Location: 7.81°S. lat 26.00°W. long
Impact Weight: 30,836 lb.
Lunar Orbit & Events:
Initial Apocynthion/Pericynthion (LOI): 169 x 58.4
Descent Orbit (DOI) 58.8 x 9.6 NM
CSM Circularization: 63.9 x 56.0 NM
Landing Site Coordinates: 3.66°S lat 17.48°W. long
Lunar Landing Time: 4:18 a.m. EST, February 5, 1971
LM Liftoff from Lunar Surface: 1:49 p.m. EST, February 6, 1971
Ascent Stage Impact on Lunar Surface: 7:46 p.m. EST, Feb 6, 1971

APOLLO 14

Ascent Stage Impact Coordinates: 3.42°S. latitude 19.66° W. longitude
Ascent Stage Impact Velocity: 5500 fps
Ascent Stage Impact Weight: 5067 lb.
Mission Duration: 216 hrs 01 mins 57 secs
Time of Landing: 4:05 p.m. EST, February 9, 1971

SPACE VEHICLE AND PRELAUNCH DATA

Spacecraft delivered to KSC: Command/Service Module: November 1969
Lunar Module: Nov 1969
Launch Vehicle delivered to KSC:
First Stage (S-IC) Jan 1970
Second Stage (S-II) Jan 1970
Third Stage (S-IVB) Jan 1970
Instrument Unit (IU) May 1970
Space Vehicle Weight at Liftoff: 6,420,491 lb.
Weight Placed in Earth Orbit: 302,626 lb.
Weight Placed in Lunar Orbit: 71,702 lb.

MISSION PRIMARY OBJECTIVES
(All Accomplished)
1. Perform selenological inspection, survey, and sampling of materials in a pre-selected region of the Fra Mauro Formation.
2. Deploy and activate ALSEP.
3. Develop man's capability to work

in the lunar environment.

4. Obtain photographs of candidate exploration sites.

DETAILED OBJECTIVES AND EXPERIMENTS LAUNCH VEHICLE

(Both Accomplished)

* Impact the expended S- IVB/IU on the lunar surface under nominal flight profile conditions.

*Post-flight determination of actual S-IVB/IU point of impact within 5 km, and time of impact within 1 second.

SPACECRAFT AND LUNAR SURFACE

1. Contingency Sample Collection (Accomplished)

2. Apollo Lunar Surface Experiment Package (Apollo 14 ALSEP) which included the following: (Accomplished)

a. Lunar Passive Seismology (S- 031)

b. Lunar Active Seismology (S-033)

c. Suprathermal Ion Detector (S-036)

d. Low Energy Solar Wind (S- 038)

e. Cold Cathode Ionization Gauge (S- 058)

f. Lunar Dust Detector (M-515)

3. Lunar Geology Investigation (S-059) (Accomplished)

4. Photographs of Candidate Exploration Sites (Accomplished)

5. Laser Ranging Retro-Reflector (S-078) (Accomplished)

6. Soil Mechanics (S-200) (Accomplished)

7. Portable Magnetometer (S-198) (Accomplished)

8. Visibility at High Sun Angles (Accomplished)

9. Mobile Equipment Transporter Evaluation (Accomplished)

10. Selenodetic Reference Point Update (Accomplished)

11. Bistatic Radar (S-170) (Accomplished)

12. CSM Orbital Photographic Tasks (Accomplished)

13. Assessment of EVA Operation Limits (Accomplished)

14. CSM Oxygen Flow Rate (Accomplished)

15. Solar Wind Composition (S-080) (Accomplished)

16. Thermal Coating Degradation (Accomplished)

17. EVA Communication System Performance (Not Accomplished)

18. Gegenschein From Lunar Orbit (S- 178) (Accomplished)

19. S-Band Transponder (Accomplished)

IN-FLIGHT DEMONSTRATIONS

(All Accomplished)

Electrophoretic Separation Heat Flow and Convection Liquid Transfer Composite Casting

UNUSUAL FEATURES OF THE MISSION

1. First SPS DOI maneuver.

2. First use of the mobile equipment transporter.

3. Longest total EVA time to date.

4. Longest distance yet traveled on the lunar surface.

5. Largest weight of lunar samples returned to date.

6. First use of short rendezvous.

7. First use of the In-flight Demonstrations

Significant Spacecraft differences from Apollo 13:

Command/Service Module

* The SM cryogenic oxygen tanks

were redesigned to remove the fans; to eliminate, as far as possible, flammable materials; to improve the design for fabrication and assembly; and to replace teflon insulated conductors with stainless steel sheathed conductors.

* A third cryogenic oxygen tank with its associated piping was added in SM bay 1 to provide backup to existing two tanks.

* A solenoid isolation valve was added to isolate the third oxygen tank from the other two.

* An auxiliary battery was added in SM bay 4 to provide electrical Power backup if fuel cell Power should become unavailable.

* Water bags having a 40# capacity were added to provide return enhancement for the CSM water system.

Lunar Module

* Anti-slosh baffles were added to the descent stage propellant tanks to improve PQGS flight performance and decrease propellant level uncertainty.

* Wiring was added to enhance Power transfer capability from LM ascent stage to CSM.

* Modifications were made to the LM batteries to prevent any free KOH from causing short circuits.

* Modifications were made to descent stage Quads I and II structure to provide for stowage of laser ranging retro reflector and the lunar portable magnetometer.

Significant launch vehicle (POGO). from Apollo 13:

S-II Stage

* A center engine LOX feed-line accumulator was added to alleviate potential 16 Hz structural/ propul-

sion oscillations (POGO).

* A backup center engine cutoff system was provided to eliminate possibility of high g loads developing to destructive levels.

* Two Position mixture ratio control valves were incorporated to simplify propellant mixture control system by eliminating the Interface with the IU computer.

S-IVB Stage

* Two position mixture control valves were incorporated to simplify the propellant mixture control system.

Other significant configuration changes from Apollo 13:

Crew Systems

* The buddy secondary life support System (BSLSS) was incorporated to provide capability to supply cooling water to an astronaut with a failed portable life support System (PLSS) from a working PLSS.

RECOVERY DATA

Recovery Area: Mid-Pacific Ocean

Landing Coordinates: 27°0'S., 172°39'30"W.(Stable I)

Recovery Ship: USS New Orleans

Crew Recovery Time: 4:53 p.m. EST, February 9, 1971

Spacecraft Recovery Time: 5:55 p.m. EST, February 9, 1971

REMARKS

Apollo 14 was launched at 4:03 p.m. EST on January 31, 1971 after an unscheduled 40 minute hold occurred at T-8 minutes and 2 seconds, due to high overcast clouds and rain. All launch vehicle systems performed satisfactorily throughout the expected lifetime. Following orbital insertion, all major systems were verified, preparations were completed and the S-IVB second burn was carried out as planned to insert the spacecraft into a translunar trajectory. Difficulties were encountered in the docking of the CSM and LM and a successful "hard dock" was not accomplished until the sixth attempt. Other aspects of the translunar journey were nominal and only one midcourse correction was made. The S-IVB stage impacted the moon's surface, as planned. The Apollo 12 passive seismometer located 169 km northwest of the impact point recorded the event 37 seconds later. LM separation and descent were as planned and it was reported that the LM landed on an 8 degree slope about 30 to 60 feet short of the planned target in the Fra Mauro area. Minor communications difficulties delayed the start of the first extra vehicular activity (EVA) period 49 minutes. During EVA-1, the Apollo lunar surface experiments package (ALSEP) was deployed approximately 500 feet west of the LM and the laser ranging retroreflector an additional 100 feet west of the ALSEP. The laser ranging team at the MacDonald Observatory in Texas reported high quality "returns" from the retroreflector shortly after deployment. All ALSEP experiments are now functioning as expected. EVA-1 was terminated after 4 hours and 49 minutes. Following a rest period, the second EVA was started 2 hours and 27 minutes ahead of schedule. The LM Crew set out on a geology traverse, using the mobile equipment transporter (MET), to carry tools, cameras, and the lunar Portable magnetometer. Lunar samples were also collected. During the geology 'traverse, various samples, Photographs and terrain descriptions were obtained. Two measurements were made with the Portable magnetometer to determine variations in the moon's magnetic field. Difficulty encountered in traversing the rough terrain resulted in the furthermost point of the traverse being established short of the rim of Cone Crater in order to allow sufficient time for completing all mandatory scientific tasks in EVA-2. EVA-2 was terminated after a total of 4 hours 28 minutes. Approximately 169 pounds of samples were collected, and the total traverse distance for the two EVA's was 3.3 km. During the LM lunar surface stay various astronomic and lunar photographic tasks were performed from the CSM in lunar orbit. Ascent of the LM from the lunar surface, rendezvous and docking with the CSM were performed as planned. No docking problems were encountered but the docking probe was brought back to earth for post flight analysis. The LM ascent stage was impacted on the moon and signals were recorded by both the Apollo 12 and Apollo 14 ALSEPS. During the return flight from the moon four inflight technical demonstrations of equipment and processes designed to illustrate the use of the unique condition of zero-gravity in space were performed. Only one midcourse correction was required during the transearth flight. The CM and SM separation, reentry and splashdown were carried out according to plan. The CM landed in the Pacific Ocean approximately 675 miles south of Samoa and about 4 nautical miles from the prime recovery ship USS New Orleans.

APOLLO 15 (AS-510) FLIGHT SUMMARY

GENERAL
Spacecraft: CM-112, SM-112, LM-10
Launch Vehicle: SA-510
Launch Complex: 39A
Flight Crew:
Commander (CDR) David R. Scott
Command Module Pilot (CMP) Alfred M. Worden
Lunar Module Pilot (LMP) James B. Irwin
Launch Time: 9:34 a.m. EDT, July 26, 1971
Launch Azimuth: 80.088°
Earth Orbit: 91.5 x 92.5 NM
S-IVB/IU Lunar Impact Time: 4:59 p.m. EDT, July 29, 1971
Velocity of Impact: 8455 fps.
Lunar Location: 1°S. lat, 11.87°W. long
Impact Weight: 30,786 lb.
Lunar Orbits and events:
Initial Apocynthion/Pericynthion (LOI): 170x58 NM
Descent Orbit (DOI): 58.5 x 9.2 NM
DOI Trim: 59.9 x 9.6 NM
CSM Circularization: 64.7 x 53 NM
Landing Site Coordinates: 26°05'N. lat, 3°39'E. long
Lunar Landing Time: 6:16 p.m. EDT, July 30, 1971
LM Liftoff from Lunar Surface: 1:11 p.m. EDT, August 2, 1971
Ascent Stage Impact on

APOLLO 15

Lunar Surface: 11:04 p.m. EDT, August 2, 1971
Ascent Stage Impact Coordinates: 26.22' N. latitude 0°15'E. longitude
Ascent Stage Impact Velocity: 5562 fps
Ascent Stage Impact Weight: 5259 lb.
Subsatellite Launch: 4:13 p.m., EDT, August 4, 1971; 76.3 x 55.1 NM, 28.7° inclination
Mission Duration: 295 hrs 11 mins 53 secs
Time of Landing: 4:46 p.m. EDT, August 7, 1971

SPACE VEHICLE AND PRELAUNCH DATA

Spacecraft delivered to KSC:
Command/Service Module: January 1971
Lunar Module: Nov 1970
Lunar Roving Vehicle: Mar 1971
Launch Vehicle Delivered to KSC:
First Stage (S-IC): July 1970
Second Stage (S-II): May 1970
Third Stage (S-IVB): June 1970
Instrument Unit (IU): June 1970
Space Vehicle Weight at Liftoff: 6,407,758 lb.
(107,142 lb. payload)
Weight Placed in Earth Orbit: 309,330 lb.
Weight Placed in Lunar Orbit: 74,522 lb.

MISSION PRIMARY OBJECTIVES

(All Accomplished)
1. Perform selenological inspection, survey, and sampling of materials and surface features in' a pre-selected area of the Hadley-Apennine region.
2. Emplace and activate surface experiments.
3. Evaluate the capability of the Apollo equipment to provide extended lunar surface stay time, increased EVA operations, and surface mobility.
4. Conduct in-flight experiments and photographic tasks from lunar orbit.

DETAILED OBJECTIVES AND EXPERIMENTS LAUNCH VEHICLE

(Both Accomplished)

* Impact the extended S-IVB/IU on the lunar surface under nominal flight profile conditions.
* Post-flight determination of actual S-IVB/IU point of impact within 5 km, and the time of impact within one second.

LUNAR SURFACE

(All Accomplished)
1. Contingency Sample Collection
2. Documented Sample Collection (Apennine Front)*
3. Apollo Lunar Surface Experiment Package (Apollo 15 ALSEP), which included the following:
a. Lunar Passive Seismology (S-031)
b. Lunar Tri-Axis Magnetometer (S- 034)
c. Medium Energy Solar Wind (S- 035)
d. Suprathermal Ion Detector (S- 036)
e. Cold Cathode Ionization Gauge (S-058)
f. Lunar Heat Flow (S-037)
g. Lunar Dust Detector (M-515)
4. Drill Core Sample Collection*
5. Laser Ranging Retro-Reflector (S-078)

6. Lunar Geology Investigation (S-059)
7. LRV Evaluation
8. EVA Communications with LCRU/GCTA
9. EMU Assessment of Lunar Surface
10. LM Landing Effects Evaluation
11. Solar Wind Composition (S-080)
12. Soil Mechanics (S-200)

* Part of Lunar Geology Investigation (S-059)

IN-FLIGHT

(All accomplished except noted)
1. Gamma-Ray Spectrometer (S-160)
2. X-Ray Flourescence (S- 161)
3. SM Orbital Photographic Tasks
a. 24" Panoramic Camera
b. 3" Mapping Camera
c. Laser Altimeter (Partially Accomplished)
4. Subsatellite
a. S-Band Transponder (S-164)
b. Particle Shadows/Boundary Layer (S-173)
c. Magnetometer (S-174)
5. Bistatic Radar (S-170)
6. S-Band Transponder (CSM/LM) (S-164)
7. Alpha-Particle Spectrometer (S-162)
8. Mass Spectrometer (S-165)
9. UV Photography - Earth and Moon (S-177)
10. Gegenschein from Lunar Orbit (S-178) (Partially Accomplished)
11. CM Photographic Tasks
12. SIM Thermal Data (Accomplished)
13. SIM Bay Inspection During EVA
14. SIM Door Jettison Evaluation
15. Visual Observation from Lunar Orbit
16. Visual Light Flash Phenomenon

OTHER

(All Accomplished)

* LM Descent Engine Performance
* Apollo Time and Motion Study

* Bone Mineral Measurement (M-078)
* Total Body Gamma Spectrometry (M- 079)
* Apollo Window Meteoroid (S-176)

UNUSUAL FEATURES OF THE MISSION

1. First Apollo use of 90-NM earth parking orbit.
2. First use of direct, minimum energy trajectory to the moon.
3. First use of scientific instrument module (SIM).
4. Largest spacecraft payload yet put in lunar orbit. (74,522 lb.)
5. Highest lunar orbit inclination (28.9°) during a manned mission.
6. First LM landing using 25° descent trajectory.
7. First use of stand-up EVA on the lunar surface.
8. Establishment of sensor networks by deployment of third station for the lunar passive seismometer and laser reflector experiments.
9. First use of extended capability CSM, LM, space suits, and PLSS's.
10. First use of manned lunar roving vehicle and lunar surface navigation devices.
11. First use of lunar communications relay unit and ground commanded TV assembly.
12. Longest total EVA time to date (18.6 hr.).
13. Longest distance yet traveled on the lunar surface (27.9 km).
14. Largest weight of lunar sample material returned to date (Approx. 169 lb.)
15. Deepest core sample of lunar material yet obtained. (7 ft. 6 in.).
16. First scientific exploration of lunar mountain and rille areas.
17. First TV observation of LM ascent from the lunar surface.
18. First launch of a subsatellite in lunar orbit.

19. Longest manned duration in lunar orbit (74 orbits).
20. First EVA from CM in deep space.
21. First in-flight TV and photos of moon during solar eclipse.
22. First lunar landing mission with no post-mission quarantine requirements.

Significant spacecraft differences from Apollo 14

Command/Service Module

* A third SM cryogenic H2 tank and associated plumbing were added for increased electrical power capability.
* A Scientific Instrument Module with a jettisonable door was added to bay IV of the SM, with associated controls in the CM, to increase the in-flight science capability by the operation of on-board sensors and a long-duration subsatellite in lunar orbit.
* A scientific data system was added to collect and transmit SIM experiment and camera data, with the capability for real- time data transmission simultaneously with tape recorder playback of lunar far side data.
* The CM environmental control system was modified to provide for in-flight EVA by the CMP to retrieve film from the SIM bay cameras, and external handholds and a foot restraint were also added for the EVA.

Lunar Module

* The descent stage propellant tanks were enlarged to provide for increased LM landing weight and landing point selection through longer powered descent burns.
* The descent engine specific impulse was increased by the addition of a quartz liner and a ten-inch nozzle extension.
* A GOX tank, a water tank, a descent stage battery, and a new water container were added to increase the lunar stay time to 68 hours.

* Stowage provisions were incorporated for the LRV in quad I and for the LRV-carried equipment pallet in quad III.

Crew Systems and Lunar Mobility

* Provided A7L-B spacesuit with improved mobility and durability to increase the lunar surface EVA efficiency and stay time, including increased drinking water and fruit bar provisions in the CDR and LMP suits and in-flight EVA capability for the CMP suit.
* Modified the life support system (PLSS) to increase O_2, H_2O, and LiOH quantities and battery power to increase the range and efficiency of lunar surface operations by extending the maximum EVA time to seven hours.
* Added the lunar roving vehicle to increase the range and scientific return of lunar surface traverses.
* Added a lunar communications relay unit (LCRU), carried either on the LRV or by an astronaut, to enhance uplink and down-link telemetry, voice, and TV communications during lunar surface traverses.
* Added a ground commanded TV assembly (GCTA) to provide earth-controlled color TV monitoring of lunar surface activities through the LCRU, including LM ascent and post-liftoff lunar surveys.

Significant launch vehicle changes from Apollo 14:

S-IC Stage
* Increased payload capability approximately 500 lb. by increasing the outboard engine LOX depletion time.
* Increased payload capability approximately 100 lb. by removing four of eight retro-rocket motors.

* Increased payload capability 600 lb. by reorificing the F-1 engines to provide greater thrust.

S-II Stage
* Eliminated single engine failure points and increased payload capability approximately 90 lb. by removing four ullage motors.
* Improved reliability and payload capability approximately 210 lb. by replacing LH2 and LOX ullage pressure regulators with fixed orifices.

Instrument Unit
Improved Power supply reliability by adding redundant + 28 volt Power for the ST-124 stabilized platform system.
* Modified the launch tower avoidance yaw maneuver which increased tower clearance assurance and reduced launch wind restrictions.
* Increased the accuracy of TLI burn cutoff in the event of IU platform failure by modifying the CM computer to provide backup burn cutoff capability.

RECOVERY DATA
Recovery Area: Mid-Pacific Ocean
Landing Coordinates: 26°07'N., 158°09'W. (Stable I)
Recovery Ship: USS Okinawa
Crew Recovery Time: 5:26 p.m. EDT, August 7, 1971
Spacecraft Recovery Time: 6:20 p.m. EDT, August 7, 1971

REMARKS
Apollo 15 was launched on time after an exceptionally smooth countdown. All launch vehicle systems performed nominally, except that the S-IVB J-2 engine delivered greater than predicted thrust, which had no adverse effects on the mission. TLI was performed as predicted and CSM separation, turnaround, and docking accomplished without problems. Spacecraft

separation from the S-IVB/IU/SLA was accomplished shortly thereafter. Two S-IVB APS burns were performed to accomplish the targeted S-IVB/IU lunar impact. The actual impact was 188 km northeast of the Apollo 14 site and 355 km northeast of the Apollo 12 site. The impact provided seismic data to depths of 50-100 km vs. 30 km from previous impacts. Shortly after docking, during translunar coast, both telemetry and cabin indications identified an electrical short in service propulsion system (SPS) control circuitry and troubleshooting isolated the Problem to the delta V thrust A switch or adjacent wiring. Special SPS burn procedures developed and conducted for the MCC-2 maneuver indicated that SPS Bank A could be operated satisfactorily in the manual mode for subsequent firings, all of which were performed successfully. The SIM bay door was successfully jettisoned into a heliocentric orbit 4.5 hours before LOI. The SIM experiment and cameras were initiated successfully after LOI. Because the high orbital inclination established a flight path over the major lunar mascons, the orbital decay rate was greater than anticipated. A DOI trim burn was performed with the SM RCS to change the orbit from 59 x 7.1 NM to 59.9 x 9.6 NM. CSM/LM undocking and separation were delayed 25 minutes because of a loose umbilical connector, after which the CSM "circularized" its orbit to 64.7 x 53 NM. After the LM landed at the Hadley-Apennine site, sightings performed by the Commander during his 35 minute stand-up EVA in the top hatch and sightings from the CSM fixed the landing site about 600 meters north-northwest of the target point. The first EVA traverse was conducted to the Apennine mountain front immediately after deploying the lunar roving vehicle (LRV). After the 10.3-km LRV traverse on EVA-1 the ALSEP was deployed and activated. One 150-cm probe of the Heat Flow Experiment was emplaced; however, the second probe was not completed until EVA-2 because of drilling difficulties with the battery powered Apollo Lunar Surface Drill (ALSD). All ALSEP units operated normally and good data was received. The 300-cube Laser Ranging Retro-Reflector was deployed and has been acquired with greater ease than was possible with the previous smaller (100 cube) units. EVA-1 was terminated at 6 hr. 33 min. due to higher than normal O_2 usage by the Commander, whose usage rate was normal on subsequent EVA's. The LRV traverse on EVA-2 was 12.5 km, during which speeds of 12-13 kph were achieved and excellent LRV controllability and slope-climbing capability were demonstrated. Lunar samples were collected at the Apennine front and the secondary crater complex to the south, and final station tasks were performed back at the ALSEP site. The EVA-2 duration was 7 hr. 12 min. EVA-3 featured a 5.1 km LRV traverse to the terrace area of Hadley Rille, and samples, photography, and geologic descriptions were obtained. The 2.4-meter core tube drilling was completed, which produced a core sample of 58 distinct layers of various sized soil and rock materials. The 4 hr. 50 min. EVA was completed after positioning the LRV to monitor LM ascent with the LCRU/GCTA. The ascent stage lift-off was observed on TV; however, the LCRU unexpectedly stopped responding to signals on August 3rd, before the lunar sunset and solar eclipse could be observed. The new and improved lunar surface equipment, combined with the geologic training of the crew, produced outstanding scientific achievements. The LRV averaged 9.2 kph during its 3 hr. 2 min. riding time with good navigational accuracy, yet consumed only half the expected battery power. The enhanced mobility of the spacesuits was quite evident on TV as the crew performed difficult tasks with increased dexterity. Linear patterns in the mountain slopes and the Hadley Rille wall structure were reported by the crew and extensively photographed, including 500 mm Hasselblad photographic surveys. Seventy documented samples, core tubes, trench samples, and comprehensive samples amounted to about 169 pounds of lunar material returned to earth. Of equal scientific significance was the performance of the in-flight geochemical experiments and CMP tasks during the six-day period in lunar orbit. The gamma-ray-spectrometer detected higher levels of radioactivity on the lunar far side, and lower average levels than that measured in the Fra Mauro samples. X-ray spectrometer data indicates richer abundance of aluminum in the highlands, especially on the far side, yet greater concentrations of magnesium in the maria. The alpha-particle spectrometer data indicates that radon diffusion on the moon is three orders of magnitude less than on earth. The mass spectrometer detected an unexpected population of molecules in lunar orbit. Although the velocity/height sensor was erratic, almost all of the panoramic camera's 6500 feet of film is usable as high resolution stereo photography. The mapping camera achieved excellent results with all 4700 feet of its film. Laser altimeter performance started to degrade during revolution 26 and was inoperative after revolution 38; however, initial results were

very significant in that the moon's center of mass was found to be offset. The subsatellite particles and fields sensors returned excellent initial data, including detection of a new mascon near the east limb and indications that mascons vary in their intensity. The 76.3 x 55.1 NM lunar orbit is designed to give the subsatellite a lifetime of at least one year. All CM photographic tasks were successfully accomplished except for the Gegenschein experiment. Visual observations by the CMP achieved important sightings, such as a rille within another rille, potential worthwhile landing sites, volcanic cone structures, and previously undetected details of major crater structures. During transearth coast, the CMP performed a 38-min. in-flight EVA to retrieve the panoramic and mapping film cassettes. He made a third excursion to inspect the SIM bay and to investigate the V/H malfunction, the mapping camera extend/retract mechanism failure, and the mass spectrometer boom position. TEI burn accuracy was such that no midcourse correction was required until MCC-7. CM separation and atmospheric entry were normal; however, one of three main parachutes partially collapsed during descent and a slightly harder than planned landing occurred about one NM from the planned point (285 NM north of Hawaii) and 5.5 NM from the prime recovery ship. The astronauts were flown to Hickman AFB, Hawaii the next day, and thence to Ellington AFB, Texas. Spacecraft and Crew systems performance were near nominal throughout the mission. All anomalies were rapidly analyzed and either resolved or safely dispositioned by workaround procedures developed with effective ground/flight coordination. The flight crew performance was outstanding throughout the mission.

APOLLO 16 (AS-511) FLIGHT SUMMARY

GENERAL
Spacecraft: CM-113, SM-113, LM-11
Launch Vehicle: SA-511
Launch Complex: 39A
Flight Crew:
Commander (CDR) John W. Young
Command Module Pilot (CMP) Thomas K. Mattingly
Lunar Module Pilot (LMP) Charles M. Duke
Launch Time: 12:54 pm EST, April 16, 1972
Launch Azimuth: 72°
Earth Orbit: 95 x 90 NM
S-IVB/IU Lunar Impact Time: 4:02 p.m. EST., April 19, 1972
Velocity of Impact: 8711 fps. (Est.)
Lunar Location: 2.1°N. latitude, 24.3° W. longitude (Est.)
Impact Weight: 30,805 lb.(Est.)
Lunar Orbits and events:
Initial Apocynthion/Pericynthion (LOI): 170.3 X 58.1 NM
Descent Orbit (DOI): 58.5 x 10.9 NM
Initial CSM Separation: 59.2 x 10.4 NM
CSM Circularization: 68 X 53.1 NM
Landing Site Coordinates: 9°N. latitude, 15°31'E. longitude
Lunar Landing Time: 9:24 p.m. EST, April 20, 1972

LM Liftoff from Lunar Surface:
8:26 pm EST, Apr 23 1972
Ascent Stage Jettison: 3:54 p.m. EST, April 24, 1972
Subsatellite Launch: 4:56 p.m. EST, April 24, 1972 66.6 X 52.8 NM
Mission Duration: 265 hrs 51 mins 05 secs
Time of Landing: 2:45 pm EST Apr 27 1972

SPACE VEHICLE AND PRELAUNCH DATA

Spacecraft delivered to KSC:
Command/Service Module: July 1971
Lunar Module: May 1971
Lunar Roving Vehicle: September 1971

Launch Vehicle Delivered to KSC:
First Stage (S-IC): Sept 1971
Second Stage (S-II): July 1971
Third Stage (S-IVB): July 1971
Instrument Unit (IU) : Sept 1971

Space Vehicle Weight at Liftoff: 6,439,605 lb.
(107,158 lb. payload)
Weight Placed in Earth Orbit: 308,734 lb.
Weight Placed in Lunar Orbit: 76,109 lb.

APOLLO 16

MISSION PRIMARY OBJECTIVES
(All Accomplished)
1. Perform selenological inspection, survey, and sampling of material and surface features in a pre-selected area of the Descartes region.
2. Emplace and activate surface experiments.
3. Conduct in-flight experiments and photographic tasks.

DETAILED OBJECTIVES AND EXPERIMENTS
LAUNCH VEHICLE
(Partially Accomplished)

* Impact the expended S-IVB/IU in a pre-selected zone on the lunar surface under nominal flight profile conditions to simulate the ALSEP passive seismometers.
* Post-flight determination of actual S-IVB/IU point of impact within 5 km, and the time of impact within one second.

LUNAR SURFACE
(All accomplished except noted)
1. Documented Sample Collection*

Apollo Lunar Surface Experiment Package
which included the following:
2. Lunar Heat Flow (S-037) (Not Accomplished)
3. Lunar Tri-Axis Magnetometer (S-034)
4. Lunar Passive Seismology (S-031)
5. Lunar Active Seismology (S-033)
6. Drill Core Sample Collection (Part of Lunar Geology Investigation S- 059)
7. Lunar Geology Investigation (S-059)
8. Far UV Camera/Spectroscope (S-201)
9. Solar Wind Composition (S-080)
10. Soil Mechanics (S-200)
11. Portable Magnetometer (S-198)
12. Cosmic Ray Detector (Sheets) (S-152)
13. Lunar Roving Vehicle Evaluation

IN-FLIGHT
(All accomplished except noted)
1. Gamma-Ray Spectrometer (S-160)
2. X-Ray Fluorescence (S- 161)
3. SM Orbital Photographic Tasks
a. 24" Panoramic Camera
b. 3" Mapping Camera
c. Laser Altimeter
4. Subsatellite (Partially Accomplished)*
a. S-Band Transponder (S-164)
b. Particle Shadows/Boundary Layer (S-173)
c. Magnetometer (S-174)
5. S-Band Transponder (CSM/LM)(S-164)
6. Alpha-Particle Spectrometer (S-162)
7. Mass Spectrometer (S- 165)
8. UV Photography - Earth and Moon (S-177)
9. Gegenschein from Lunar Orbit (S-178)
10. Visual Light Flash Phenomenon
11. Microbial Response in Space Environment (M-191)
12. CM Photographic Tasks
13. Visual Observations from Lunar Orbit
14. Bistatic Radar (S-170)
15. Skylab Contamination Study (Partially Accomplished)
16. Improved Gas/Water Separator
17. Body Fluid Balance Analysis
18. Subsatellite Tracking for Autonomous Navigation (Not Accomplished)
19. Improved Fecal Collection Bag
20. Skylab Food Package

* The CSM shaping burn prior to subsatellite ejection was not performed, as described under Remarks. As a consequence, the sub-satellite's orbit was such that it impacted the Lunar surface on May 30 after a number of low altitude passes. All experiments performed as planned and significant low altitude data was acquired during the vehicle's life.

OTHER
(All Accomplished)
*Voice and Data Relay
* Apollo Time and Motion Study Bone Mineral Measurement (M-078)
* Apollo Window Meteoroid (S- 176)
* Biostack (M-211)

IN-FLIGHT DEMONSTRATION
* Electrophoretic Separation (Accomplished)

UNUSUAL FEATURES OF THE MISSION
1. Largest spacecraft payload yet put in lunar orbit.
2. First scientific exploration of lunar highlands and Cayley formation.
3. First use of the moon as an astronomical observatory.
4. Longest total lunar surface EVA time to date (20.3 hr.).
5. Largest weight of lunar samples returned to date Appx. 213 lb
6. Longest lunar stay time to date (71 hr. 2 min.).

Significant spacecraft differences from Apollo 15:

Command/Service Module
* The time delay in the RCS control box was increased from 42 seconds to 61 seconds for mode IA aborts to reduce possible land landing hazards with pressurized propellant tanks.
* Installed transparent Teflon shields to strengthen meter glass and to retain glass particles in case of breakage.
* Installed Inconel parachute links in place of nickel plated links to reduce probability of parachute riser link failures due to flaws.
* Replaced selected early series switches with 400 series switches to reduce the possibility of switch failure.

Lunar Module
* Descent stage batteries were improved to prevent case cracking and to increase electrical capacity.
* Added glycol shutoff valve to increase battery temperature, if required, to maximize electrical capacity.
* Added an exterior glass doubler to the range/range rate meter window to reduce stress. Added tape and particle shield as required to other meters.

Apollo 16 SLA
* Changed ordnance adhesive in pyro train to avoid a lead acetate reaction. Crew Systems and Lunar Mobility
* New LRV Beat belts were installed to eliminate adjustment and latching problems.
* Stage fittings in the pressure garment assembly were modified to provide greater mobility and reliability, and gloves were reinforced for greater wearability. Lunar Surface Equipment
* The ground commanded TV assembly incorporated new clutch assemblies, a new elevation drive motor, and temperature control modifications to preclude previous flight problems.

Significant launch vehicle changes from Apollo 15:

S-IC Stage
* Four retro-rocket motors were added (8 total) to improve S-IC/S-II separation characteristics.

S-II Stage
* Structure was modified to increase safety factors and to improve POGO stability.
* Several single-point relay failure modes were eliminated in the engine start/cutoff circuitry.

S-IVB Stage
* Fuel and LOX feedline bellows were changed from stainless steel to 2-ply solar duct.

Instrument Unit
* The LVDC was modified to distinguish between failures of upper and lower engines for proper abort guidance programming.
* Redesigned command decoder by adding solder joint stress relief to eliminate solder joint cracks for improved reliability.

RECOVERY DATA

Recovery Area: Mid-Pacific Ocean
Landing Coordinates: 0°43'S., 156°13'W. (Stable II)
Recovery Ship: USS Ticonderoga
Crew Recovery Time: 3:20 p.m. EST, April 27, 1972
Spacecraft Recovery Time: 4:45 p.m. EST, April 27, 1972

REMARKS

Apollo 16 was launched on time after a countdown with no unscheduled holds. All launch vehicle systems performed nominally in achieving an earth parking orbit of 95 x 90 NM. A nominal translunar injection (TLI) burn was performed after one and a half orbits. During CSM/LM docking, particles were noticed coming from the area of a LM close-out panel. The crew entered the LM early, at 8:17 GET, to determine system status. All systems were normal, and it was later determined that the particles were flakes of thermal protection paint, the loss of which would have no adverse effect on LM operations. The first S-IVB APS burn for lunar impact was nominal. Because of APS module No. 1 helium depletion due to external leakage, the APS-2 maneuver was not performed. Tracking of the S-IVB/IU was lost at 27:10 GET due to signal loss from the IU command and communications system. Lunar impact was detected by the Apollo 12, 14, and 15 seismometers and was approximated at 75:08 GET and 260 km NE of the targeted impact point. Spacecraft operations were close to-nominal until the CSM Prepared for the SPS circularization burn on the lunar far side. A problem was detected in the secondary yaw actuator servo loop which drives the SPS gimbal in backup mode. The burn was not performed as scheduled and the LM PDI burn on Rev. 13 was delayed. The CSM maneuvered to station-keeping position with the LM while trouble shooting was performed. Analysis concluded that the secondary system was operable and the landing could proceed. To minimize the remaining SPS engine firings, lunar orbit plane change 2 and the subsatellite shaping burn were deleted. Subsequently, it was decided to shorten the mission one day. Circularization was performed on Rev. 15, and LM PDI was accomplished on Rev. 16. The landing in the Descartes area was only 230 meters NW of the planned target point. Because of the almost 6-hour delay in landing caused by the SPS control problem, EVA-1 was rescheduled to follow a full crew rest period. Before performing the traverse to Flag Crater, the crew deployed and activated the Apollo lunar surface experiments package (ALSEP) and other experiments. During ALSEP deployment, the Commander inadvertently pulled the heat flow experiment cable loose at its central station connector and that experiment was abandoned. Approximately 42 pounds of samples were collected during the 7-hr. 11-min. EVA and total distance traveled by the LRV was 4.2 km. The second 11.4-km traverse took the crew about half way up 500- meter high Stone Mountain, 4.1-km south of the LM. The lunar roving vehicle provided excellent mobility and stability, achieving eleven to fourteen kilometers per hour (kph) over rocky, pockmarked surfaces and easily climbing to 15- to 20-degree slopes at about 7 to 8 kph. Extensive sampling was accomplished, and about 71 pounds were collected during the 7-hr. 23-min. EVA. The extension of the 7-hr. EVA was possible because PLSS consumables usage was lower than predicted. The EVA-3 duration of 5 hr. 40 min. was judged adequate to meet objectives while holding the ascent and rendezvous work day to an acceptable length. The LRV traverse was 4.5 km to North Ray Crater, the biggest yet explored on an Apollo mission. Very interesting house rocks were sampled, one about house-size, another with permanent shadowed area in the lee of the sun line and interesting "drill-holes" normal to its surface. Polarimetric photography was accomplished and additional portable magnetometer readings were obtained. At one point during the downslope return to the LM the LRV recorded about 18 kph. Approximately 100 pounds of samples were collected during the 11.4- km traverse. The film cassette from the far UV camera was retrieved after 51 hours recording 11 planned celestial targets. The 71-hour stay in the Descartes area featured excellent experiment, LRV, TV, and crew systems operation; revised theories of Cayley formation; less evidence of volcanism than expected, and the highest recordings of local magnetic field of any Apollo landing site. 1809 frames of 70 mm film and 4 1/2 magazines of 16 mm film were exposed during the 20-hour 15-minute total EVA time. One hundred eleven documented samples totaled approximately 213 pounds. LM ascent, rendezvous, and docking were normal. However, after jettison from the CSM the LM ascent stage lost attitude control and began tumbling at about 3° per second, probably because of an open circuit breaker in the primary guidance and navigation system, and it could not be deorbited as planned. The ascent stage is expected to stay in lunar orbit approximately a year before impacting the surface. Lunar orbital science and photographic tasks were successfully conducted throughout most of the 64 CSM orbits. The subsatellite was launched 4 hr. 20 min. before transearth injection; however, because of the decision not to perform the orbit shaping burn its lifetime was much shorter than the planned one year. To maintain the orbital time line after the delayed CSM circularization event, a GET Clock update of 11 min. 48 sec. was made at 118:06:31. To minimize checklist changes during transearth coast, another GET adjustment of 24 hr. 34 min. 12 sec. was made at 202:25, after the transearth injection maneuver. The spacecraft was depressurized for 1 hr. 23 min. during trans-earth coast for the CMP's EVA to retrieve mapping and panoramic camera film cassettes. He also inspected the SIM bay to report on experiment conditions, and the microbial response in space environment was conducted for 10 minutes outside the open hatch. Two Small midcourse corrections were made during transearth coast. Final detailed objectives were completed, and an 18-min. TV press conference was conducted. CM separation, entry, and descent were normal, with water landing 0.3 from the target point and 3.5 NM from the primary recovery ship (PRS). The CM was righted from the stable II position, and the Crew was greeted aboard the PRS 35 minutes later. The crew's health was excellent throughout the flight. Because of the in-flight arrhythmias experienced by the Apollo 15 crew, special pre-flight procedures, in-flight dietary Supplements, and longer scheduled rest periods were instituted for the Apollo 16 crew. The post-flight adaptation periods were less than those experienced after Apollo 15. Numerous "glitches" and System anomalies were rapidly analyzed by the Support/flight controller/crew team and were effectively resolved to minimize the mission impact.

APOLLO 17 (AS-512) FLIGHT SUMMARY

GENERAL

Spacecraft: CM-114, SM-114, LM- 12
Launch Vehicle: SA-512
Launch Complex: 39A
Flight Crew:
Commander (CDR) Eugene A. Cernan
Command Module Pilot (CMP) Ronald E. Evans
Lunar Module Pilot (LMP) Harrison H. Schmitt
Launch Time: 12:33 a.m. EST, December 7, 1972
Launch Azimuth: 91.5°
Earth Orbit: 92.5 X 91.2 NM
S-IVB/IU Lunar Impact Time: 3:27pm EST Dec 10 1972
Velocity of Impact: 2545 meters per second
Lunar Location: 4°12'S. latitude, 12°18'W. longitude
Impact Weight: 30,712 lb.
Lunar Orbits and Events:
Initial Apocynthion/ Pericynthion (LOI): 170.0 X 52.6 NM
Descent Orbit:
DOI-1: 59 X 14.5 NM
DOI-2: 59.6 X 6.2 NM
Initial CSM Separation: 61.5 X 11.5 NM
CSM Circularization: 70 X 54 NM
Landing Site Coordinates: 20°12.6'N. lat, 30°45.0'E. long
Lunar Landing Time: 2:55 pm EST, Dec 11, 1972
LM Liftoff from Lunar Surface: 5:55 p.m. EST, Dec 14, 1972
Ascent Stage Jettison: 11:31 a.m. EST, Dec 15, 1972

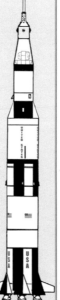
APOLLO 17

Mission Duration: 301 hrs 52 mins
Time of Landing: 2:25pm EST Dec 19 1972

SPACE VEHICLE AND PRELAUNCH DATA

Spacecraft Delivered to KSC:
Command/Service Module: March 1972
Lunar Module: June 1971
Lunar Roving Vehicle: June 1972
Launch Vehicle Delivered to KSC:
First Stage (S-IC): May 1972
Second Stage (S-II): October 1970
Third Stage (S-IVB): December 1970
Instrument Unit (IU) June 1972
Space Vehicle Weight at Liftoff: 6,445,127 lb.
Weight Placed in Earth Orbit: 311,151 lb.
Weight Placed in Lunar Orbit: 76,540 lb.

MISSION PRIMARY OBJECTIVES

(All Accomplished)
1. Perform selenological inspection, survey, and sampling of material and surface features in a pre-selected area of the Taurus-Littrow region.
2. Emplace and activate surface experiments.
3. Conduct in-flight experiments and photographic tasks.

DETAILED OBJECTIVES AND EXPERIMENTS

LAUNCH VEHICLE
(All Accomplished)

* Impact the expended S-IVB/IU in a pre-selected zone on the lunar surface under nominal flight profile conditions to stimulate the ALSEP passive seismometers.

* Post-flight determination of actual S-IVB/IU point of impact within 5 km, and the time of impact within one second.

LUNAR SURFACE

(All accomplished)
1. Documented Sample Collection*

Apollo Lunar Surface Experiment Package
which included the following:
2. Lunar Heat Flow (S-037)
3. Lunar Seismic Profiling (S-203)
4. Lunar Surface Gravimeter (S-207) (Operating, constrained mode)
5. Lunar Atmospheric Composition (S-205)
6. Lunar Ejecta and Meteorites (S-202)
7. Lunar Geology Investigation (S-059)
8. Drill Core Sample*
9. Surface Electrical Properties (S-204)
10. Lunar Neutron Probe (S-229)
11. Traverse Gravimeter (S-199)
12. Cosmic Ray Detector (S-153)

* Part of Lunar Geology Investigation (S-059)

IN-FLIGHT
(All accomplished)
1. Lunar Sounder (S-209)
2. SM Orbital Photographic Tasks
   a. 24" Panoramic Camera
   b. 3" Mapping Camera
   c. Laser Altimeter
3. IR Scanning Radiometer (S-171)
4. Far UV Spectrometer (S-169)
5. S-band Transponder (CSM/LM) (S-164)
6. Visual Light Flash Phenomenon
7. CM Photographic Tasks
8. Visual Observations from Lunar Orbit
9. Food Compatibility Assessment
10. Protective Pressure Garment
11. Skylab Contamination Study

OTHER
(All Accomplished)
* Gamma-Ray Spectrometer (S-160)
* Apollo Window Meteoroid (S-176)
* Soil Mechanics (S-200)
* Biostack II A (M-211)
* Biocore (M-212)
* Long Term Lunar Surface Exposure Tasks

IN-FLIGHT DEMONSTRATION
Heat Flow and Convection
(Accomplished)

UNUSUAL FEATURES OF THE MISSION
1. Longest single lunar surface EVA time (7.6 hr.)
2. Longest total lunar surface EVA time (22 hr. 5 min.)
3. Longest lunar distance traversed with LRV on one EVA (19 km).
4. Longest total distance traversed with LRV (35 km).
5. Largest weight of lunar sample material returned (Approximately 250 lbs.)
6. Longest lunar stay time (75 hrs.)
7. Longest time in lunar orbit (147 hr. 48 min.).
8. Longest Apollo mission (301 hr. 52 min.)

Significant spacecraft differences from Apollo 16:

Command/Service Module
* Retractable HF antenna booms and a UHF antenna were added for the Lunar Sounder experiment.
* The SIM bay was modified to incorporate the Lunar Sounder, IR Scanning Radiometer, and Far UV Spectrometer experiments.

RECOVERY DATA
Recovery Area: Mid-Pacific Ocean
Landing Coordinates: 17°52'S., 166°09'W.
Recovery Ship: USS Ticonderoga
Crew Recovery Time: 3:17 p.m. EST, December 19, 1972
Spacecraft Recovery Time: 4:28 p.m. EST, December 19, 1972

REMARKS
Apollo 17 was launched at 12:33 a.m. EST. on December 7, 1972 after an unscheduled hold of 2 hours 40 minutes. The launch countdown proceeded smoothly until T-30 seconds at which time an automatic cutoff occurred. The hold was caused when the Terminal Countdown Sequencer (TSC) failed to command pressurization of the S-IVB LOX tank. Subsequent to a recycle and hold at T-22 minutes an additional hold was called at T-8 minutes. The duration of the holds delayed the launch by 2 hours and 40 minutes. All launch vehicle systems performed nominally in achieving an earth parking orbit of 92.5 X 91.2 NM. A nominal translunar injection (TLI) burn was performed over the Atlantic Ocean during the third orbit. Nominal S-IVB APS burns and LOX dumps targeted the S-IVB/IU for lunar impact. Impact occurred at 89:39 GET. Actual impact coordinates were 4°12'S. and

12°18'W., about 160 km northwest of the planned target point. The event was recorded by the Apollo 12, 14, 15, and 16 Apollo Lunar Surface Experiment Packages (ALSEPs). The CSM separated from the S-IVB/IU/LM at 3:42 GET and docked with the LM at 3:57 GET. A docking ring latch problem required LM pressurization and hatch removal for troubleshooting. Following latch inspection the CSM/LM combination was successfully ejected at 4:45 GET. The spacecraft trajectory was nominal with only one midcourse correction (MCC-2) required. At 65:00 hours GET, a clock update of 2 hours and 40 minutes was performed to put the mission back on the original schedule and all subsequent maneuvers were performed at the nominal times. The landing in the Taurus-Littrow area was at 20°12.6'N. and 30°45.0'E., within 200 meters of the planned point. Extra-Vehicular Activity (EVA-1) commenced at 117:01:36 GET and terminated at 124:13:47 for a total duration of 7 hrs. 12 min. 11 sec. After deployment the Lunar Roving Vehicle (LRV) and prior to traversing to the ALSEP site, the CDR inadvertently knocked the right rear fender extension off of the LRV fender. The fender extension was subsequently secured to the fender with tape. The ALSEP and the Cosmic Ray experiment were deployed. Steno Crater was sampled as Station 1A in lieu of the pre-planned station (Emory Crater). The new Station was selected because of the accumulated delay in the EVA for completion of the ALSEP deployment. During the traverse to Station 1A, the fender extension came off and as a result, the crew and LRV experienced a great deal of dust. The Surface Electrical Properties

Transmitter was deployed near the end of the EVA. Since the crew did not get far enough out to deploy the 3 pound Explosive Package, only the 1/2 pound and 1 pound charges were deployed on EVA-1. EVA-2 started at 140:34:48 GET, approximately 1 hour, 20 minutes late, and ended at 148:12:10 GET. Total EVA time was 7 hours, 37 minutes, 22 seconds. Prior to starting the EVA traverse, the crew received instructions from the ground controllers for improvising a replacement for the lost fender extension. A rig of 4 chronopaque maps, taped together and held in position by two clamps from portable utility lights, made an excellent substitute for the extension and the crew did not experience the dust problem as on EVA-1. Stations 2 (Nansen), 3 (Lara), 4 (Shorty), and 5 (Camelot) were visited according to pre-mission plan although station times were modified. During the traverse, the crew deployed the 1/8 pound, 6 pound and 1/4 pound explosive packages and an orange colored material, believed to be of volcanic origin, was found at Station 4. The LMP revisited the ALSEP site at the end of the EVA in order to verify the Lunar Surface Gravimeter was properly deployed and leveled. Total distance covered was approximately 19 km. EVA-3 was initiated at 163:32:35 GET about 50 minutes late, and was terminated 7 hours and 15 minutes 31 seconds later at 170:48:06 GET. Exploration of the stations was modified during the traverse. Photographs and documented samples were obtained at all stations. About 145 pounds of Samples were retrieved, and the LRV traversed a total of 11.6 km. The 3 pound explosive package, left over from EVA-1, was deployed in addition to 1/4

pound and 1/8 pound charges. The total time for the three EVAS was 22 hours, 5 minutes, 4 seconds. The total distance traveled in the lunar rover was about 35 km. The combined weight of Samples was about 115 kg (250 pounds), plus 2 double cores and 1 deep drill core. After 75 hours on the lunar Surface the LM ascent stage lifted off and performed a nominal rendezvous and docking. Following crew and equipment transfer to the CSM, the LM ascent stage jettison and CSM separation were completed as planned. Ascent Stage Deorbit was initiated and lunar Surface impact occurred at 195:57 GET. The event was observed by the four Apollo 17 geophones and the Apollo 12, 14, 15 and 16 ALSEPs. The seismic experiment explosive packages were detonated as planned and were recorded by the geophones; however, only one detonation was viewed by the ground controlled television before an over-temperature failure in the Lunar Communications Relay Unit occurred. Apollo 17 At the end of the 75th lunar orbit a nominal Transearth Injection maneuver was performed. During transearth coast the command module pilot performed an EVA to retrieve film from the lunar sounder experiment and the mapping and panoramic cameras in three trips to the SIM bay. The EVA required a total of 1 hour and 7 minutes. One small midcourse correction was made prior to entry interface. The CM landed in a stable 1 position 0.4 NM from the planned point and about 4.3 NM from the prime recovery ship the USS Ticonderoga. The crew was picked up by helicopter and was on board the Ticonderoga 52 minutes after splashdown.

A dramatic launch tower shot of Apollo 12 showing the rain and bad weather at lift off. November 14th 1969

The all-Navy crew of Apollo 12, Charles "Pete" Conrad, Richard Gordon, Alan Bean (l to r)

Mission controllers watch a telecast from Apollo 13 just minutes before the near-fatal explosion aboard the craft.

Command Module Pilot Jack Swigert assists in the assembly of a make-shift air-scrubbing device.

Despite desperately harsh conditions aboard, the Apollo 13 crew still took time to take photographs, including this one of a beautiful half-Earth.

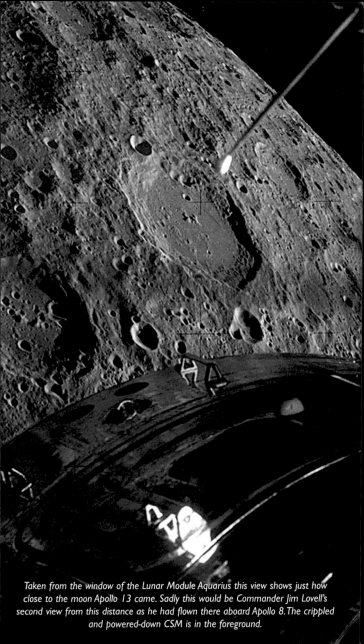

Taken from the window of the Lunar Module Aquarius this view shows just how close to the moon Apollo 13 came. Sadly this would be Commander Jim Lovell's second view from this distance as he had flown there aboard Apollo 8. The crippled and powered-down CSM is in the foreground.

The make-shift air-scrubber can be seen strapped together from duct-tape, a piece of cardboard and a sock. This impromptu device, conceived by controllers on the ground, saved the crews' lives by removing poisonous CO_2 from the depleted atmospheric reserves of the cabin.

When the oxygen tank (above) exploded it ripped away one entire section of the Service Module as shown in this picture taken by the crew just before re-entry.

Apollo 13 breaks through the clouds over the Pacific ocean to the intense relief of everyone. President Nixon was on hand to greet the crew on their safe return to Earth. (inset)

The Apollo 13 command module Odyssey is hauled aboard the recovery ship. The multiple redundancies in Apollo proved their worth and saved the crew.

After nine months of extensive modifications Apollo returned to space on January 31st 1971. The Apollo 14 crew consisted of Stuart Roosa, Alan Shepard and Edgar Mitchell (right). Flight destinations were changed and crews altered due to the failure of Apollo 13 and the perception in Washington that further flights were a waste of money.

In this picture of the ALSEP the infamous mortar can be seen in the foreground. It remains unfired to this day.

Shepard and Mitchell were able to walk much further than previous crews and still do useful science thanks to the MET. In the picture inset at left they struggle up the slope of Cone Crater.

Mitchell selects an appropriate tool. (both false color)

This image does much to demonstrate how far away Mitchell and Shepard walked during their EVA. Antares is crystal clear but potentially dangerously far away. In case of mishaps the crew had a new oxygen "buddy" system as a back-up.

Shepard makes his way back to Antares (left, false color)

The lunar lift off was filmed by Edgar Mitchell with the 16mm camera. Debris is scattered by the engine exhaust.

Apollo 15 was launched on July 26th 1971. The Saturn booster was pushed to its design llimit to accomodate more equipment, including the first lunar rover.

The Apollo 15 crew was composed of Commander David Scott (left) and two astronauts who had not flown in space, Al Worden (centre) and James Irwin (right)

The starkly beautiful Hadley Rille, the target for Apollo 15, was located at a much higher latitude than other Apollo landings, requiring more fuel and a more potentially hazardous flight-plan.

The Apennine mountains make a stupendous backdrop for the Apollo 15 lunar module and rover. The two-seater vehicle (bottom right) served the crew flawlessly over three days taking them almost 28 kilometers. In what has become one of the iconic pictures of the Apollo program, the 10,000 foot high Mount Hadley Delta looms above the site of mankind's fourth lunar base (opposite left)

Commander Dave Scott reorients the high-gain antenna on the lunar rover to enable the transmission of TV pictures back to Earth. This image is a false color composite of three photographs taken by Jim Irwin at the base of Mount Hadley Delta that can be seen rising in the background. Distance is deceiving in the absence of atmosphere so the two mile hike to the summit appears deceptively close. Inset at left is an image of the hammer and feather dropped by Scott during his demonstration of Galileo's theories on gravity.

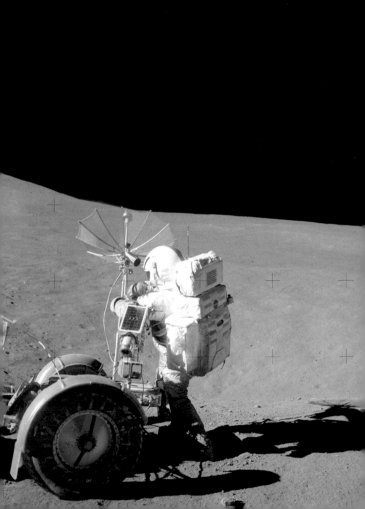

The Apollo 15 service module was equipped with the first of three scientific instrument modules (SIM). Pilot Al Worden spent three days mapping the moon in unprecedented detail from orbit (above) before retrieving the precious film cartridges in a space walk on the way home (artist's concept below).

April 16th 1972 the Saturn V lofts almost six and a half million pounds away from the launch pad. The fifth moon landing mission would be crewed by Thomas Mattingly, John Young and Charles Duke (l to r inset left)

John Young cannot contain his exuberance as he jumps clear of the ground while saluting the flag.

The CSM, Casper, dwindles in the distance as the Earth rises.

Another view of Casper as the LM, Orion, pulls away 10 miles above the lunar surface.

Charles Duke poses with the flag at the Descartes landing site. The second lunar rover can be seen behind the flag.

Apollo 16 was the fifth lunar landing mission and studying lunar geology was one of the paramount reasons for the mission. A large boulder is examined at left.

←

The lunar module Orion is the only spark of color in an otherwise desolate landscape. The RCA color television camera is in the foreground.

↓

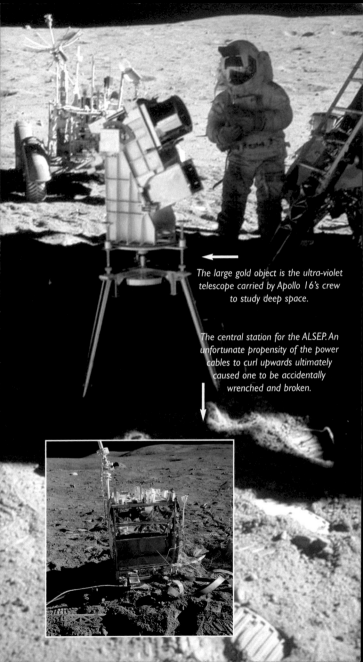

The large gold object is the ultra-violet telescope carried by Apollo 16's crew to study deep space.

The central station for the ALSEP. An unfortunate propensity of the power cables to curl upwards ultimately caused one to be accidentally wrenched and broken.

Once again the lunar rover proved invaluable as it allowed Young and Duke to cover great distances.

The LM was left behind as the astronauts sought out volcanic rocks.

Carrying more equipment than ever before Apollo 16 was an almost flawless mission. Spacesuits had been refined as can be seen in this picture, although the addition of a potable drinking supply caused some consternation inside the helmets.

The LM ascent stage prepares to dock in lunar orbit.

The angle of the sunlight could make profound differences in the color of the lunar soil. Looking into the sun in this image the soil appears almost white.

The crew of Apollo 17 was Harrison Schmitt, Ron Evans and Eugene Cernan (sitting). Only Cernan had flown in space previously.

The last manned mission to the moon leaves on December 7th 1972. The final voyage involved place 150 tons of spacecraft and cargo into space. The only night launch of a Saturn V was witnessed by throngs of people eager to catch one last glimpse of man's most powerful machine taking flight.

The Taurus-Littrow valley had originally been selected as the landing site for Apollo 14. It was a rich geological target and proved to be a tremendous choice for Harrison Schmitt, the only geologist to walk on the moon. As can be seen in this picture the two explorers left many tracks during their three day sojourn.

Commander Eugene Cernan takes a moment to salute the US flag.

Harrison Schmitt works with part of the ALSEP
in this beautiful panorama compiled from two
separate images. A spectacular image of the full
Earth showing Africa is inset above.

Challenger was the third LM to carry a fully equipped LRV to the moon.

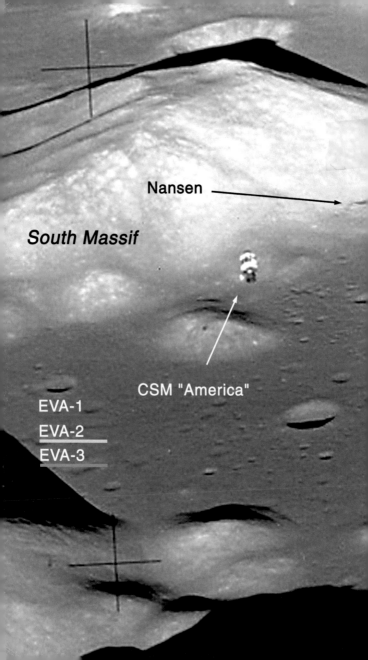

Nansen

South Massif

CSM "America"

EVA-1
EVA-2
EVA-3

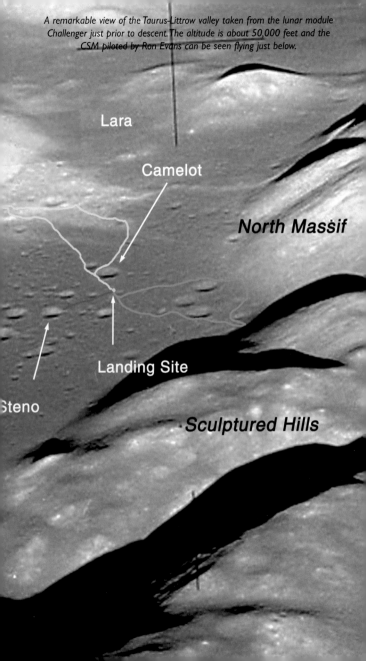

A remarkable view of the Taurus-Littrow valley taken from the lunar module Challenger just prior to descent. The altitude is about 50,000 feet and the CSM piloted by Ron Evans can be seen flying just below.

Lara

Camelot

North Massif

Landing Site

Steno

Sculptured Hills

Cernan poses with the Sculptured Hills behind him.

America boasted an impressive array of scientific equipment in its SIM bay.

The LRV lost a fender which was promptly replaced by a folded map.